Macrologistics℠ MANAGEMENT

A Catalyst for Organizational Change

The St. Lucie Press/APICS Series on Resource Management

Titles in the Series

Supply Chain Management: The Basics and Beyond
by William C. Copacino

Applying Manufacturing Execution Systems
by Michael McClellan

Macrologistics Management: A Catalyst for Organizational Change
by Martin Stein and Frank Voehl

Macrologistics℠ MANAGEMENT

A Catalyst for Organizational Change

Martin Stein
Frank Voehl

CRC Press
Taylor & Francis Group
Boca Raton London New York

CRC Press is an imprint of the
Taylor & Francis Group, an **informa** business

Library of Congress Cataloging-in-Publication Data

Macrologistics management: a catalyst for organizational change/ Martin Stein and Frank Voehl.
p. cm.
(St. Lucie Press/APICS series in resource management)
Includes bibliographical references and index.
ISBN 1-88401-539-5
1. Business management. I. Title. I. Series.
RP103.E6R54 1997
872.5'.25146—dc21
for Library of Congress

97-51272
CIP

International Standard Book Number 1-88401-539-5
Library of Congress Card Number 97-51272

Dedication

To Marshall McDonald
1918–1997

The godfather of the Macrologistics Management program at FPL.
A man of boundless energy and vision, he brought to corporate America
a new management system that was to have profound effects upon thousands
of organizations across this country and world that he loved so much.
Thank you for teaching us how to learn.

To my three sons, Chris, Jeff and Greg.
Thank you for teaching me how to love.
And to Micki for teaching me how to be loved.

Frank Voehl

To my wife and family, whom I love so much.
Beatriz, Joshua and Giovanni, Michelle and Jacqueline
Thank you for filling my life with joy.

Martin Stein

Contents

Preface xi

SECTION I: OVERVIEW

1 **Macrologistics Management: The Catalyst for Organizational Change** 3

2 **The Macrologistics Systems Model: Putting the Pieces Together** 15

SECTION II: ALIGNMENT

3 **Policy Deployment and QFD: The Twin-Engines of Macrologistics** 27
Profile 3.1: FPL 39
Profile 3.2: 3M 50

4 **Suply Chain Management** 55
Ralph Lewis
Profile 4.1: Becton Dickinson 69
Profile 4.2: Nippon Steel 74
Profile 4.3: Microage 77
Profile 4.4: Value Chain 10-Point Model 78

5 **Information Technology** 81
Profile 5.1: Levi Strauss 97
Profile 5.2: How Technology Wins 98
Profile 5.3: Achieving Logistics Quality 99
Profile 5.4: KAO 100

Profile 5.5: Better Information 101
Profile 5.6: Data Scatter 102
Profile 5.7: Livonia 103
Profile 5.8: ELM 105

SECTION III: MOBILIZATION

6 **Total Innovative Management** 109
Profile 6.1: Burlington Railroad 121
Profile 6.2: Rubbermaid 130
Profile 6.3: Hitachi 132
Profile 6.4: Eastman Chemical 134
Profile 6.5: Samsung 136
Profile 6.6: Stanco 137
Profile 6.7: Ikea 138
Profile 6.8: Shisiedo 140
Profile 6.9: IBM 142
Profile 6.10: Chrysler 143
Profile 6.11: Norfolk Southern 145

7 **Just-In-Time (JIT)/JIT II™** 149
Profile 7.1: Bose 160
Profile 7.2: A Day in the Life 167
Profile 7.3: Bell Labs 169
Profile 7.4: JIT and the Management Accountant 170
Profile 7.5: Inventory Policy and Practice 171
Profile 7.6: Impact of JIT Systems on Small Businesses 172

8 **ValueStream Quality System** 175
Jeffrey Vengrow and Frank Voehl
Profile 8.1: Whirlpool 188
Profile 8.2: Hewlett Packard 190
Profile 8.3: Return on Quality 191
Profile 8.4: Texaco 192
Profile 8.5: Kietretsu in America 193

SECTION IV: INTEGRATION

9 **Benchmarking Using Critical Success Factors** 195
Profile 9.1: General Electric 208
Profile 9.2: Motorola 209

Profile 9.3: Heroes on the Help Desk 210
Profile 9.4: APQC 212

10 **Process Management and Reengineering** 215
Profile 10.1: Xerox 225
Profile 10.2: IBM Restructures 227

11 **The Balanced Scorecard Corporate Measurement System** 231
Profile 11.1: Mars 240
Profile 11.2: Air Products 242
Profile 11.3: Total Cost: A New Trend 243

12 **Implementing Macrologistics Management Strategies: A Blueprint for Change** 245

Glossary 253

Index 267

Preface

It is during the turbulent times at sea that those who serve on watch at the lighthouse must project a clear, reliable guiding light to the ships around them — ships struggling against changing, dangerous tides, crashing waves, and the ever-present danger of running aground. The promise of this work is to act as a beacon to those organizations in today's global marketplace who are struggling against the tide and need a new approach to logistics management.

Macrologistics can help your organization identify breakthrough strategies that have the potential for significant impact on the overall organizational performance and competitive success. The benefits of such a strategy — speed, stretch, and boundarylessness — are within the potential of the breakthrough change strategies outlined in this book. The objective of the change is to have a major impact upon corporate performance.

Logistics is like plumbing: nobody pays any attention to it until it breaks down. And most organizations do not have information systems that adequately support the logistics process, so they are unable to effectively manage it. Also, most logistics improvements are "micro," and they target incremental cost savings in limited tactical areas, such as warehousing, transportation costs, or inventory management. Worse yet, management often fails to appreciate the full potential of logistics as a strategic tool, while logistics policies are either inconsistent or invisible.

This work is about the use of logistics strategies to change the overall functioning of processes and to alter the fundamental manner in which an organization carries out its logistics practices and processes. While we have found that Macrologistics strategies can be effectively deployed in the implementation of a structured change process, we have also found that no one organization has implemented the complete package, as can be seen from a reading of the many Profiles following Chapters 3 to 11.

In the Profiles we are indebted to James Higgins who graciously granted permission to incorporate selected profiles from his collection, which are independently published in his work *"Innovate or Evaporate"*. These Profiles are intended to keep readers abreast of new trends and stories in the field of Macologistics. They are also intended to help readers benefit from the insights and experiences of experts and practitioners, of which Higgins is in the field of the Management of Innovation. (Contact James Higgins at the Higgins Innovation Center at 407-647-5344.)

The ever-increasing complexity of the challenges facing organizations, along with the pace of change, signal the escalating pressure that will be brought to bear on logistics professionals to either play provocative, strategic partnership roles or be left behind as marginal contributors. Also, the challenge to develop World-Class work culture that integrates logistics quality and strategic concerns is imminent. Organizations that cannot or will not develop the coordinated integration of strategic management, quality management, information systems management, and logistics management will, in the long run and the short run, become ineffective global competitors.

More and more, customers expect rapid customization of products and services, and genuine responsiveness to their changing needs in order to sustain their purchasing commitment. Macrologistics Management offers the logistics professional an opportunity to take a more proactive partnership role with senior management, and to review their policies and practices to insure that the highest quality of services is being rendered and continuously improved to the benefit of the internal and external customers.

We trust that you will find this book both useful and beneficial. We wish those readers who are attempting to implement Macrologistics Management maximum successes and minimal roadblocks and constraints. For further advice and counsel in this area, feel free to contact the authors through the publisher.

The Authors

Martin M. Stein is an internationally recognized expert in the design of corporate change processes and programs. He has developed these programs to enhance customer satisfaction, leadership, loyalty, and market penetration. He developed the term "Macrologistics" while serving as Deputy Director of the Center for Transportation Studies at the Massachusetts Institute of Technology. He used this innovative approach to design a benchmarking conference with the American Productivity and Quality Center. At this conference over 20 company case studies on the use of logistics to develop competitive breakthrough strategies were presented.

This book represents a major milestone in Dr. Stein's career because it captures the dynamic elements of change present in a wide variety of challenging business process reengineering and business transformation consulting assignments. These projects were completed for companies such as General Motors, Toyota, the U.S. Department of Defense, and Chrysler. Dr. Stein is currently the founder of a consulting firm that specializes in providing support to organizations that desire to utilize advanced management techniques to achieve breakthrough changes in customer loyalty, customer service and market penetration. He also supports the Bose Corporation's JIT II Education and Research Center as Program Director. He is also affiliated with major international consulting organizations such as the Gemini Consulting Company and the Forum Corporation.

Martin Stein is currently developing innovative concepts for change integration, information systems management and "adaptive engineering." The latter process serves as a mechanism for problem solving and continuous improvement that merges features of process reengineering developed by Michael Manning with adaptive learning techniques developed by Peter

Senge of MIT. Dr. Stein is also designing an updated implementation approach for JIT II, a new methodology for supplier partnering created by Lance Dixon of Bose Corporation and currently used by companies such as IBM, AT&T, Honeywell and Intel.

Dr. Stein holds a BS and MA in Economics from the University of Maryland, College Park, and a D.E.S. and Doctorate of Science from the University of Paris VI (Sorbonne). His international background has led to consulting and teaching assignments around the world for leading automotive manufacturers, airlines, railroads, and a variety of other organizations including the United Nations. He is the Founder and President of several professional associations such as the Society of Automotive Analysts and is listed in Who's Who in Finance and Industry and Who's Who in the World.

Frank Voehl is President and Chief Executive Officer of Strategy Associates, Inc., and former Corporate Vice President and General Manager of Qualtec Quality Services, Inc., an FPL Group company. Since 1986, Frank has been responsible for overseeing the implementation of Quality Management Systems with organizations in such diverse industries as telecommunications and utilities, federal, state and local government agencies, public administration and safety, pharmaceuticals, insurance/banking, manufacturing, and institutes of higher learning.

Frank's consulting specialty is in the areas of Quality Management and related logistics disciplines, including strategic planning, productivity and Macrologistics management, as well as the softer areas of team building, synergy, and group dynamics. As a result of his work during the past twenty years, new systems and operating models have been created for continuous quality improvement, workforce staffing, corporate measurement, Quality Management implementation, and Quality Council operations. His projects have also included Cost-of-Quality measurement, Integrated Zero Based Budgeting, ISO 9000 evaluations, as well as synergy and group dynamics for high performance teams.

In 1991, he formed Strategy Associates, which serves as a consulting network specializing in Quality Management, continuous improvement, and teamwork. During the past ten years, his he has served numerous manufacturing and service industry clients, both nationally and abroad, with a focus on measurement systems that are effective, economical, and long-lasting. From 1992 to 1995, Frank helped implement a National Quality Award for the Bahamas, during which time he served on the Board of Judges, training examiners and performing numerous site visits and application reviews.

In 1986 he helped establish the Qualitec Quality Services consulting and training arm of FPL, where he served as founding General Manager and Chief Operating Officer until November 1991. Under his leadership, Qualtec Quality Services grew from four professionals in 1986 to over one hundred in six years, achieving a position among the top ten U.S. Quality Management consulting organizations, as noted by Business Week in October 1991. During the past fifteen years, Frank has consulted with hundreds of organizations, including such Fortune 500 companies as AT&T, Eli Lilly, Michigan Consolidated Gas, Chase Manhattan Bank, New York Life, ConEd, McDermott, and others. He has also served various organizations in diverse environments including Xerox, Proctor and Gamble, GOJO Industries, the Internal Revenue Service, city and state governments and Departments of Transportation, Community Quality Councils, universities, and others, both in this country and abroad.

During 1990 and 1991, Frank was appointed Chancellor of the Institute for Competitive Advantage and was involved with the establishment of the Quality Institute at the University of Miami, where he was a lecturer and visiting professor in the School of Industrial Engineering from 1983 to 1989. In 1991 he was appointed as the Quality Representative for the Internal Revenue Service twenty-member Commissioner's Advisory Group (CAG), and has served on Strategic Planning and Executive Committees of the AQP and ASQC. In 1994 and 1995, Frank also served as CEO of the Anro Metals Manufacturing Company, a Florida-based sheet metal fabrication plant.

During this period he was a pioneer Lead Facilitator and one of the original designers of FPL's award-winning Quality Management Program. In this capacity he served from 1981 to 1985 on the FPL Corporate Design/Development Team, whose work led to the formation of the Malcolm Baldrige National Quality Award. During the 1980s, his work with the Florida Department of Transportation resulted in a savings of approximately $150 million for the taxpayers of the State of Florida, in addition to a statewide process improvement of the forced relocation planning process.

Frank received a Bachelor of Science degree in Industrial Relations and attended Law School at St. John's University, followed by postgraduate studies at Columbia University. He is Series Editor of ten books in the field of Total Quality, including Marketing, R&D, Supplier Quality, Human Resource Management, Information Systems, History of Total Quality, and Higher Education. He is the author of hundreds of journal articles and technical presentations and co-author of a number of diverse management books such as: *ISO 9000: An Implementation Guide for Small to Mid-Sized Businesses;*

Deming: The Way We Knew Him; An Executive Guide to Implementing Quality Systems; Total Quality in Information Systems, Team Building: A Structured Approach, Problem Solving for Results, and Macrologistics Management.

About APICS

APICS, The Educational Society for Resource Management, is an international, not-for-profit organization offering a full range of programs and materials focusing on individual and organizational education, standards of excellence, and integrated resource management topics. These resources, developed under the direction of integrated resource management experts, are available at local, regional, and national levels. Since 1957, hundreds of thousands of professionals have relied on APICS as a source for educational products and services.

- **APICS Certification Programs** — APICS offers two internationally recognized certification programs, Certified in Production and Inventory Management (CPIM) and Certified in Integrated Resource Management (CIRM), known around the world as standards of professional competence in business and manufacturing.
- ***APICS Educational Materials Catalog*** — This catalog contains books, courseware, proceedings, reprints, training materials, and videos developed by industry experts and available to members at a discount.
- ***APICS — The Performance Advantage*** — This monthly, four-color magazine addresses the educational and resource management needs of manufacturing professionals.
- ***APICS Business Outlook Index*** — Designed to take economic analysis a step beyond current surveys, the index is a monthly manufacturing-based survey report based on confidential production, sales, and inventory data from APICS-related companies.
- **Chapters** — APICS' more than 270 chapters provide leadership, learning, and networking opportunities at the local level.

- **Educational Opportunities** — Held around the country, APICS' International Conference and Exhibition, workshops, and symposia offer you numerous opportunities to learn from your peers and management experts.
- **Employment Referral Program** — A cost-effective way to reach a targeted network of resource management professionals, this program pairs qualified job candidates with interested companies.
- **SIGs** — These member groups develop specialized educational programs and resources for seven specific industry and interest areas.
- **Web Site** — The APICS web site at http://www.apics.org enables you to explore the wide range of information available on APICS' membership, certification, and educational offerings.
- **Member Services** — Members enjoy a dedicated inquiry service, insurance, a retirement plan, and more.

For more information on APICS programs, services, or membership, call APICS Customer Service at (800) 444-2742 or (703) 237-8344 or visit http://www.apics.org on the World Wide Web.

OVERVIEW

I

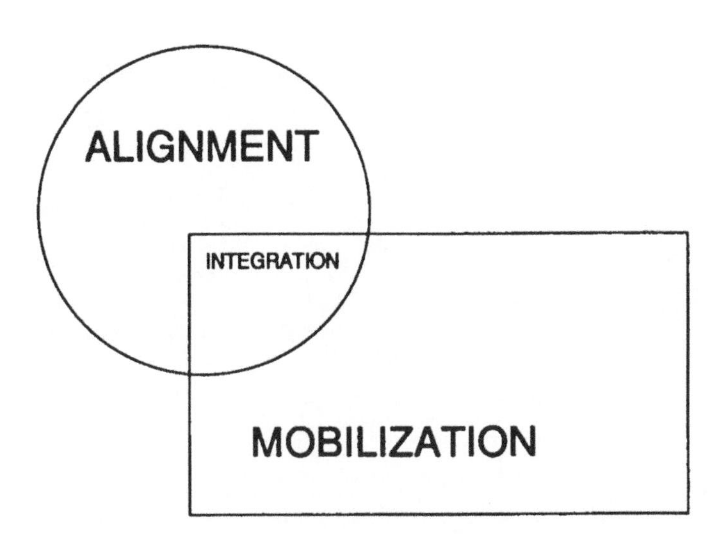

1 Macrologistics* Strategy: The Catalyst for Organizational Change

Imagine that you are the CEO of a Fortune 500 company. You have just left a board meeting where the members of the Board expressed their dissatisfaction with progress achieved last year. In your head you know that there is a rational explanation for the failure of the organization to change; in your heart you feel that there are insurmountable obstacles. The company culture or the group sense of motivation is at its lowest level in years. Unless you can demonstrate substantial change, your next task might be faxing your resume. You keep your resume in a secret compartment in the left-hand drawer of your desk. You can see yourself filling out the transmittal slip with the name of the headhunter you hired last year. You wonder if he will help, since the job he had to find a new Executive VP never turned up anyone that could help run the company and the position is still vacant. Your future is on the line and so is the company's.

In the past, you thought that Total Quality Management was a flag that everyone would salute. After a fairly successful launch and some interesting meetings with consultants — including invoices with lots of zeroes — the initial enthusiasm has waned. There is little acceptance of this process by union employees. Many line managers have complained that there are few significant changes resulting from the process. You vaguely remember a *Wall Street Journal* article describing the case where the CEO of one company achieved the holy grail of TQM — the Malcolm Baldridge Award. Now he is

*Macrologistics is a service mark of Martin Stein Associates based in Needhan, Massachusetts.

looking for a job. Wasn't he one of the candidates for that vacant Executive VP slot and didn't the Board turn him down? You are sure that was an omen.

The TQM process was informative although somewhat shallow, and you are now aware of several approaches that Japanese executives use to reduce production defects. They also use the approaches to generate teamwork for solving manufacturing glitches and to iron out customer service snafus. None of these ideas has resulted in the catalyst for change that will help transform your company. There have been no breakthrough strategies. The only person who knows what that means or thinks he does is the Human Resources VP who thinks she deserves a medal for using the word "downsizing" at every meeting she attends.

Without a major breakthrough, your competitor's market share will continue to expand. With TQM the rate of decline is slower this year but as one Board member said, that just prolongs the day when our once dominant market share numbers convert into second place. The impact of even the slow rate of decline has been horrendous. Your stock options are now almost worthless.

Imitating the Japanese has proven to be an elusive goal. You want to leapfrog competition not emulate them. You have observed that the institutional mechanisms in Japan are not the same as in your country. Given that very interesting visit to Nagoya last year, you know that is may take years to fix all of the culture differences that exist. You may never have workers who volunteer their spare time to generate two thousand implemented suggestions a year as they do at Toyota. Values take generations to change and you really only have the next 12 months, at best.

You also participated in cross-functional quality team meetings. These discussions took days of involvement from your top people and the changes suggested had a minimal effect on the big picture. Obvious changes helped to cut costs — they really should have automated the billing department two years ago. There is not much motivation to keep meetings interesting. The word in the local union hall is that the TQM process is a synonym for downsizing. Since the payment process was automated, front-line workers lost jobs and there has been little change in the numbers of rejected product. The new equipment keeps breaking down and your supervisor suspects a new form of industrial sabotage.

Diagnosis of this problem was so tough you hired a management consultant. He said there is a disconnect between the company's systems and desired behavioral change. Whenever values and systems are not synchronized, stress occurs that evolves into negative attitudes, resentment and a poor organizational climate. In time, the culture of the company blocks all change because of these fears and perceptions. New initiatives are difficult to sell. The consultant says converting a negative climate to a positive one is an arduous task. In the past, external threats such as foreign competition was used as the catalyst for change. Yet once the organization realized that the threat was not one of life or death, the old systems took over and the negative climate returns.

TQM as a catalyst for change suffered from the lack of implementation strategies. New quick and easy solutions — the low-hanging fruit — were adopted, a stall occurred that can take years to remedy. What is now needed is a new form of catalyst for change that galvanizes external stimuli, i.e., customer satisfaction with internal forces, i.e., order fulfillment. When these forces are combined, change that is substantive can be achieved.

The framework for change needs to be one where learning and continuous improvement elements of Quality Management are accepted. It is one where continuous monitoring is facilitated by a relevant and responsive information system. Creativity needed for breakthrough change is reinforced when the invisible shackles of old values embedded in outdated software systems are surfaced and eradicated. When managers are rewarded for adapting to change requirements and for achieving stretch goals, companies will be able to accomplish breakthrough goals and become World-Class competitors.

The purpose of this book is to define the term Macrologistics as a means for designing a catalyst for change that is more than superficial. By defining goals that are integrative and "macro" in nature, the kinds of solutions that emerge will yield tangible change and transform systems. The complete transformation of a business is accomplished using a systematic process for managing change and by using carefully prioritized change management strategies. When theory and experience come together, great things can happen.

The book contains a conceptual overview, along with specific case studies of companies having applied these strategies and a review of the steps needed to apply these concepts and strategies in your organization.

Macrologistics Strategy: A New Framework for Change

The CEO of General Electric, John Welch, recently stated that he expects his managers to move their company to a new type of flat organization. Flat means less vertical hierarchy that slows decisions and bureaucratizes everything. He wants to see the scope of an individual manager's efforts to be "boundaryless."

This new form of "empowered" manager will have his success measured in terms of the "speed" by which change is accomplished. Goals are acceptable only if they "stretch" the company's capabilities.

This is a wonderful confirmation of the need for macrologistics strategy. It will take stretch to accomplish significant change. Competitors who prosper will be able to meet stretch goals with speed. They will be able to accelerate the change process and implement stretch goals in a seamless fashion across organizational divisions.

In the past, economists and accountants who are largely responsible for design of corporate performance standards have focused on tangibles. The accountant wants performance to be measured in terms of tangible flows such as income and expenses. Success is defined as the condition where there is more income than expense. Economists define several forms of utility; i.e., time, place and form. What they can really measure is form utility. Microeconomics has a complete theoretical framework and computational algorithms for measuring production costs and revenues associated with the production form of utility.

Performance measures that incorporate time and place utility in addition to form utility, will make all the difference in the future. Accountants are now actively redefining their measures to calculate the "value-added" form process. This is a relatively new basis of accounting called activity-based accounting. Economists will need to define new measures for organizations who succeed by delivering more than form utility. When radical improvements, breakthroughs, in time and place utility occur, it can result in the redefinition of an industry. For example, when Federal Express, perhaps the pioneer of Macrologistics, introduced an overnight standard, they redefined the small-package delivery industry. Domino's Pizza delivers the same type of pizza as competitors but the delivery is to your home and with a standard of performance of thirty minutes.

Stretch Goals for Added Value

Stretch goals for speed will be added to the requirement for delivery of goods and services. The extra value added in terms of time reduction will be so large that new measures will be needed. Achievement of these time and place goals, such as in the new standard for manufacturing — Just-in-Time — will require huge new investments. How can that new investment be rationalized without better performance measures? Information systems and other support systems will have to adapt to these new stretch goals. What is needed is a new framework for measuring and tracking the success of the organization in delivering on these types of speed and location specific goals.

The new framework needs to cut across the organizational functions and needs to also consider the role of suppliers and vendors of materials. Further, since environmental regulations require organizations to dispose or recycle materials used in the manufacturing process. these elements also are important in designing the acceptable framework.

Although managers have used logistics in the military to fine-tune the process of delivery that is time and location specific, there is a lag in the incorporation of military logistics requirements in the civilian sector. Usually when the term logistics is used, the purpose is to reduce transportation costs by adding automated handling equipment or adding warehousing information systems.

Recently the military used logistics in a broader sense. In the Desert Storm war, General Schwartzkopf asked why helicopters had an assumed downtime of 50%. This was a constraint to the capacity to move materiel to the frontline and this assumption alone jeopardized the success of the war. Without more rapid turnaround times on helicopters, rapid deployment of troops in Kuwait could not be achieved. No one could give him a reason for the assumption, so he changed the assumption to 20% — a stretch goal. After the downtime expectation was reduced, new maintenance and parts prepositioning systems were needed and modifications were made to repair facilities. The new goal was achieved with much less system pain than had been expected.

Managers given this stretch goal, were forced to rethink the logistics chain and repair processes were rescheduled. Like the tire changing teams at the Indy 500, the new repair crews for helicopters used better teamwork to speed up the repair process. The results were extremely successful. Not just for the war effort, but also as a way to demonstrate to military managers that overall logistics goals, at the "theater" or systems wide level, could be used as a catalyst for change.

Managers using logistics strategy to change organizational processes, depart from the routine application of logistics. To differentiate between these applications of strategy, let's use the term "Macrologistics." Now we emphasize that our objectives are system wide in nature. The definition of Macrologistics is the use of logistics policies to change the overall functioning of organizational processes and to alter the fundamental way that an organization performs its logistics operations.

Once logistics is defined in this way, several benefits emerge. For one, the logistics manager is asked to refocus his or her efforts to the achievement of "stretch" goals. Today this is difficult for several reasons. Executives often consider logistics "plumbing" and the logistics process does not get "fixed" unless it breaks. Also, organizations have inadequate logistics policies. Companies frequently lack the information systems necessary to support logistics initiatives.

Another benefit of macrologistics strategies as an agent of change is that by definition they are "boundaryless." For example, if the strategy is supply chain management or JIT II™ , the entire organization is opened up to help meet the expectations of the customer. Suppliers insights and decision in JIT II™ are now targeted to meet external as well as internal customer requirements. These strategies imply that employees and suppliers are "empowered" to meet the needs of customers. This helps support the flat management organization approach where each manager acts for the benefit of the entire organization.

Unleashing New Sources of Synergy

A major advantage of Macrologistics strategy is that the process unleashes new sources of synergy. These synergies help promote creativity and they help motivate effective and rapid change. The five new sources of synergy are as follows:

1. **Suppliers** — Often suppliers have information, research and new technology that can help. Currently, many companies have only one-way dialogues with their suppliers. Opening up a two-way dialogue provides new ideas for change and the dynamic exchange of information improves productivity. JIT II is an example of a best practice in this area and several case studies in following chapters will describe this new approach.
2. **Inplants** — Also found in the JIT II process, inplants, on-site and empowered supplier "resource" managers, provide opportunities for concurrent engineering and new ideas on production efficiency. They also serve as trouble shooters and can help prevent problems before they stop the production process.
3. **Business Rules for Innovation** — Once the macrologistics strategy is deployed, it is necessary to identify how many rules need to be broken to make it work. Also, are employees and cross-functional

managers recommending flexibility in some rules? These may be cases where there are hidden opportunities to leverage rule changes to spark new change. For example, at Xerox the salesperson defines delivery dates. Meeting his expectations was easy since by practice — a kind of business rule — he always filled in the desired delivery dates. Managers were able to redefine the process and get the customers real needs into the form and forcing the delivery process to meet higher standards of service.

4. **Benchmarking Using Critical Success Factors** — Sending groups of managers out into other companies helps stimulate new ideas. Particularly if the mix can include diverse companies, new creative ideas and innovations are likely. Because Macrologistics strategies involve considerable new investments in information system change, benchmarking is desirable.
5. **Performance Measures** — Macrologistics strategies can only be effective if there is considerable effort spent on the design of new performance measures. The new order fulfillment process when it is benchmarked will have to have specific quantitative measures. These measures force considerable discussion about goals and objectives. This unleashes synergy across organizational lines.

Integration of Macrologistics Strategies and Quality Management

The use of a Quality Management process such as TQM is one way to start a business transformation process. This motivation to excel in quality has attracted many supporters because it was successfully used by Japanese companies to achieve global dominance in industries such as electronics and the automobile industry. The desire to attain the Malcolm Baldridge Award is one way the federal government hoped to stimulate corporate competitiveness and promote new investment. Corporate managers looked at the point scores in the award and tried to achieve high point scores without understanding how the process worked completely. Separate units and teams were given very specific goals. Failure to achieve implementation of the objectives of TQM was one of the outcomes of the blind adherence to literal interpretation of the Baldridge processes.

The failure to implement TQM as a way to transform business is due to the inability of managers to see the full impact of their actions and the failure to have a meaningful catalyst for change. Lack of empowerment results in superficial changes and the lack of a catalyst reduces the process to a series of staccato actions. Improvement to process with no unifying objective can result in minor change but substantial differences in the way business process are performed are less likely. Consequently the impact on the core business outputs are not substantive with little impact on the customer or market share.

Managers and executives who see little impact from the TQM process other than higher point totals on an arbitrary measure such as the 1000-point system in the Baldrige scoring process soon lose motivation for further change. What is needed is the use of an integrated catalyst for change that is lasting and that can permanently influence the way an organization performs its core processes and delivers its core outputs.

The use of a Macrologistics strategy such as a new order fulfillment goal (e.g., 24-hr turnaround on new orders) can take every component of an organization to new levels of motivation. By challenging the entire system, the strategy offers a rallying and unifying purpose for meetings. Cross-functional teams target their resources on specific and attainable goals. Further, the elimination of defects is only part of the discussion. With the Macrologistics strategy, the emphasis shifts to the accomplishment of process changes or improved performance in support systems. New investment may be required and there is little concern about negatives such as downsizing.

There is evidence that the overemphasis on the elimination of defects in the TQM process has earned this approach the label as a euphemism for downsizing. Union workers may not even participate when they learn about some who have had their jobs sacrificed on the altar of TQM. There are recent articles about managers who fear loss of employment and who are understandably less enthusiastic about pursuing TQM-based change processes.

Macrologistics offers the promise of new challenges, new market and new opportunities. When the challenges are achieved, the result is the avoidance of negative consequences such as market share decline. Now the goal is market expansion by opening up new opportunities for sales. For example, rapid delivery provides a larger geographic market area, thus instantly adding new customers.

The requirements for using Macrologistics strategy as a catalyst for a successful value-adding process are:

- **Conceptual Framework** — The change process needs to divided into three separate phases. Some strategies work well in one phase and others are needed in latter phases. The three phases are alignment, mobilization, and integration. For example, JIT II is an excellent tool for building alignment. The use of business rules analysis is a critical part of transitioning from alignment to mobilization. This strategy works well here because it eliminates obstacles in the rules process that inhibit achievement of goals that are buried in obscure locations such as software code or obsolete procedures and practices. Adaptive engineering works well when there is alignment and mobilization and the organization can focus on a rapidly improving approach to meeting new stretch goals.

- **Leadership** — Often overlooked, the need for competent leadership is critical. In some small companies, owners make poor leaders because their economic power stifles challenge and creativity. Developing change processes requires as a core competency the need for leaders who can effectively move organizations from one plateau to another. The leadership process requires CEO level ingredients such as a passion for change. Executives have to support this with a sense of urgency and the ability to manage resource trade-offs to support the changes. Leadership at the mid-level management level must assume that the vision of the CEO is shared and that attainable intermediate goals are designed. This level of management must be able to create discernable critical success factors and measures. Teamwork and communication skills must be sharpened. In addition, the ability to manage expectations during each state of the change process is needed and the techniques for doing this are often weak. Today managers try to get their performance evaluated a very specific levels. Meeting system-level goals are not included in incentive systems and organizations have weak linkages across organizational lines so the capability to measure contributions from different sources does not exist. These ingredients all imply the need for major new investment in the leadership development process if change is to be effectively managed.
- **Information Systems** — Strategies are often implemented with no way to measure their effectiveness. New investment should not be made without adequate tools for evaluation of their performance. Even if the costs are substantial, concurrent plans for upgrading information systems can be çritical success factors in watching the "needle" change as progress is achieved. New measures may be needed that evaluate changes in cycle time, improved decision times and tracking of costs related to reduction in performance gaps.

All of these required elements imply considerable investment. The investments are needed in technology, people and information systems. Change processes that do not have substantial new investments in all of these categories could fail. It is critical to balance resources and to plan the change process. The achievement of a Macrologistics strategy offers the promise of being able to radically change the organization. The change should occur with a concomitant resource investment that is carefully targeted to meeting customer needs and to making the change process deliver results that are important in the marketplace.

Examples of Macrologistics Strategies

Supply chain management is frequently cited as an important new catalyst for change. Because it is comprehensive and because it cuts across organizational lines, it qualifies to be categorized as a Macrologistics strategy. The definition of supply chain management is the creation of a management process for integrating decisions, plans and information systems from customer requirements through the manufacturing process to the suppliers of materials. Components of the process are purchasing, manufacturing, distribution, transportation, product handling and customer service. For some products, this process can also include recycling and disposal (also known as "reverse logistics"). The major advantage of this management approach is that because it is comprehensive, there are substantial opportunities to reassess the way that customers obtain value-added from the company. If all elements of the value chain are identified, barriers to meeting and exceeding customer expectations are identified. See the case studies of Becton Dickinson and Xerox which show the benefits of this approach (see Profiles 4.1 and 12.1).

The method of systematically evaluating how the customer obtains value from the supply chain helps develop a catalyst for change. The synergies and creativity that are unleashed have a dramatic effect on a company. The impact is that breakthrough changes emerge and core processes and core outputs are defined. The supply chain strategy is very useful at the beginning of the change process because it creates a blueprint of the way that an organization functions. This blueprint can discern ways that the organization is out of alignment with customer needs and this it is very useful in the alignment phase of the business transformation process.

Another very powerful tool for accomplishing alignment with customer needs is the JIT II™ process. This strategy unleashes substantial synergy by focusing on the suppliers' hidden pool of talent, knowledge and resources. Developed at the Bose Corporation, this practice enables empowerment of suppliers. The supplier acts as an in-house change agent and delivers new value by seamlessly integrating the materials flow process and aligning delivery schedules. The alignment of the supplier organization with the manufacturing process translates into better concurrent engineering, improved designs and more opportunities for delivering value to customers. For example, one supplier has added accessories to Bose products that make the products more useful in diverse applications. Case studies of this Macrologistics strategy are described in Chapter 7.

Once alignment has been achieved, it is possible to add to the resource mix. Some logistics providers add to capabilities by becoming third-party providers of logistics service. The use of this pool of highly specialized technology, information systems and skilled logistics managers helps a company add to resources quickly with little new investment. Capabilities to meet customer needs are enhanced. Therefore, this approach is useful at the mobilization phase of the change process.

Regardless of the strategy employed, measures of success need to be established early. Also, linkages to a continuous improvement process such as value analysis and value engineering are important as a means for institutionalizing the change process.

The benefits from use of Macrologistics range all the way from Federal Express and the creation of a new industry definition, to the use of inplants which reduce operating costs and generate new ideas for products at Bose Corporation. Xerox used supply chain management and discovered assets of $1.3 billion that could be sold and converted into immediate profits. General Schwartzkopf found that the Desert Storm War was logistically successful and Becton Dickinson uses these ideas to help deliver new standards of service in resupplying hospitals with their products. The benefits are substantial.

Where "micro" logistics can deliver a small cost reduction in trucking or warehousing, moving to the big picture approach of macrologistics can result in a major transformation of an organization. Chapter 2 below provides an overview of these applications and discusses their strengths and weaknesses.

You are the CEO and you need to choose a new strategy for change. Which one will work for you? These strategies can help you chart a course that meets the Board's demands. Maybe there is a way to save the company, keep your job and avoid the headhunters!

2 The Macrologistics Systems Model: Putting the Pieces Together

Most executives spend their efforts trying to reduce short-term costs and increase short-term profitability. They concentrate on logistics savings of 5% or less and usually develop these savings in one location without consideration of the long-term consequences of the savings. This is Micrologistics. When the scope of the change crosses organizational lines, there is a radical change in the nature of the product, or service, and these changes have the potential for developing major breakthroughs, i.e., 50% savings in cost or time, they are Macrologistics changes.

A systematic approach for identifying the opportunities for Macrologistics changes and using them to gain corporate competitive advantage is called a Macrologistics System. Most companies have adopted a partial approach to Macrologistics change and despite the fact that this book contains over 40 case studies of Macrologistics changes, there currently are no companies that have a fully integrated Macrologistics System as depicted in Figure 2.1.

One of this book's objectives is to accelerate the process by which companies consider this option for corporate development. In this chapter, we provide an overview of the components of Macrologistics Systems and a rationale for using these individual case studies as potential elements of a more comprehensive approach for the management of change and development of competitive advantage strategies.

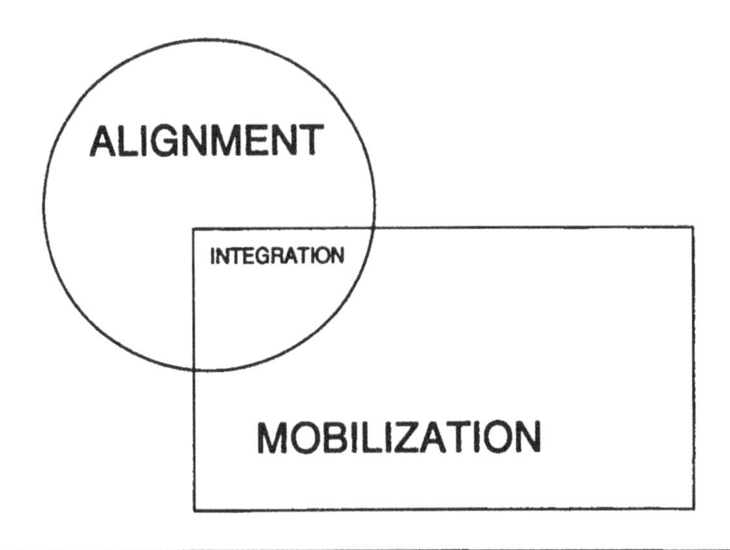

Figure 2.1 Macrologistics Management Model

Macrologistics Change Management Model

The major logic of the Macrologistics System is the change management model. There can be no development of new strategy or implementation of strategy without a logical approach to change management. Three phases of change are used in this book to structure the different element of Macrologistics. Phase I consists of Alignment activities, Phase II is the Mobilization phase and Phase III is the Integration phase, as shown in Figures 2.1 through 2.4.

Organizations whose values are misaligned will find that behavioral change is difficult. Despite the presence of a unifying vision, the vision and the related mission statements are static. What makes the vision a reality is that the organization change takes a new direction. The new direction needs to have a consensus that the direction will deliver desired results. Managers and executives that are not fully aligned will find resistance to change that dominates the strategic implementation process and often derails initiatives.

One technique described in Chapter 3 deals with the need to confirm that policy to ensure that program and practices are consistently applied. A technique for surfacing misaligned policy, program, and practices is quality function deployment. This systematic aproach checks and confirms consistency across organizational units for the implementation of quality initiatives.

Another approach that identifies the "disconnects" within the information system is business rules analysis. Cases of applying these methods, policy deployment, and QFD are presented for FPL and 3M. Business rules analysis is relatively new and applications are not yet fully developed..

Macrologistics Alignment

The need for organizations to have a logical framework for changing the supply chain has led to the creation of value chains. These chains describe the logical steps needed to deliver value to customers using materials, suppliers, and manufacturing processes. The need for every step in the chain to be identified helps locate areas where the organization does not have alignment. As a diagnostic process, the supply chain often can be used to define areas that are out of alignment. For examples of supply chain management, see the case studies provided for Becton Dickinson, Nippon Steel, and MicroAge (see Profiles 4.1, 4.2, and 4.3).

Finally, the information needed to track the supply chain and to implement strategic initiatives often implies the need for new information systems, as shown in Figure 2.2. Recently, AMOCO spent millions defining the supply chain only to find out that the new information system required to monitor the chain was going to take five more years to develop and a fortune to build. The manner in which the system is designed and how it functions, can be a critical success factor in the development of Macrologistics strategic change. Federal Express could not have offered overnight delivery without a vast investment in information systems. These systems often require alignment across organizational lines, common databases, and common definitions. Policy deployment is severely hampered by lack of consistency in the information system. This is where business rules analysis can be used to discover and fix information system disconnects that inhibit change. For example, an insurance company that was consistently losing new large accounts traced the problem back to subjective and non-competitive strategies used in underwriting new policy quotes. Once this was rectified, expansion became substantially easier to accomplish.

Information systems can be screened to determine if leverage opportunities exist. By identifying policy enablers, such as the incentive use of interest to promote sales in specific markets, alignment is enhanced and it is much easier to implement change. Case studies of information systems investments

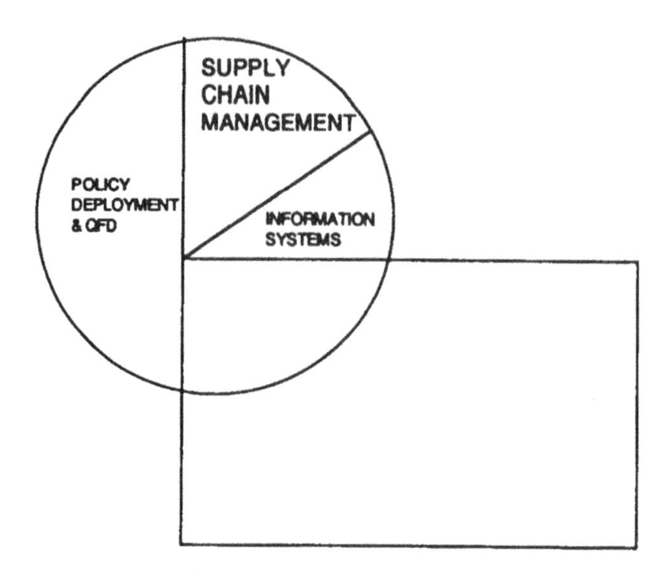

Figure 2.2 Macrologistics Management Alignment — Phase I

related to Macrologistics are provided for companies such as Levi Strauss, KAO, and ELM (see Profiles 5.1, 5.4, and 5.8).

Included in the definition of information systems is the use of new information technology such as hand-held scanners, computers and bar coding technology. Without rapid deployment of new technology, the transmission rate of data needed to support the supply chain improvement is not feasible, efficient, or effective.

Macrologistics Mobilization

Federal Express pioneered in the use of new technologies to deliver on its goals of consistency and speed. Federal Express systematically pursues continuous improvement in all of its logistics processes, is the one company that comes closest to application of Macrologistics Systems. It is no wonder that FedEx helped define a new industry and that it was the first major service company to receive the coveted Malcolm Baldrige award for Total Quality Management.

Although Federal Express embodies many of the principles of Macrologistics, they do not have a consistent change model in place, and they still have not utilized the advanced techniques of Macrologistics such as business rules analysis. For example, the recent failure to expand in the European market can be traced to the lack of a consistent approach to change and a failure to recognize the inevitable disconnects that hamper expansion in new countries and with new cultures. In addition, the techniques of Macrologistics strategies that are used by many companies, such as JIT II ™are foreign terms to FedEx. Phases II and III are Mobilization and Integration. These phases build on the platform of Alignment created in Phase I. Just as an engine needs to be aligned to provide direction for change, the engine as part of a car needs to have all of the other systems mobilized to make progress. The Mobilization phase helps companies generate more harmonious and effective use of its resources to achieve its competitive strategies.

Section III of this book summarizes several important techniques that companies can use for Mobilization. Although the companies that use Mobilization without Alignment will not be too successful, given Alignment, Mobilization is critically needed to make progress.

Chapter 6 describes the use of a comprehensive management planning process called Total Innovative Management (TIM) to orchestrate the accomplishment of new initiatives. Companies such as Burlington Railroad, Hitachi, Eastman Chemical, and Chrysler, provide illustrations of how to develop a unified approach to new strategic initiatives (see Profiles 6.1, 6.3, 6.4, and 6.10).

One important distinction is that when Japanese companies change, they have an underlying cultural consensus-building process that supports the achievement of alignment without having to reconstruct alignment. Other companies may assume alignment and try to mobilize and discover that they are not capable of achieving their goals. IBM has extremely sophisticated systems for mobilization. However, the real question for this company is are they aligned and are they integrated in their processes?

One of the more productive ways to achieve improvements in mobilization is the use of Just-in-Time principles as shown in Figure 2.3. When companies redo their inventory practices, they inevitably discover inefficiencies that need to be eliminated for progress on strategic initiatives to occur. In the case of Japanese companies such as Toyota, they re-deploy all their resources including supplier resources in an integrated fashion to achieve breakthroughs in product cycle reductions and in cost savings due to the elimination of inventory expense.

Once the JIT process savings are obtained, the question remains on how to continuously upgrade this process. One innovative, American approach for improving on the manner in which JIT is conducted involves the use of supplier partnering. This was developed at Bose Corporation and is called

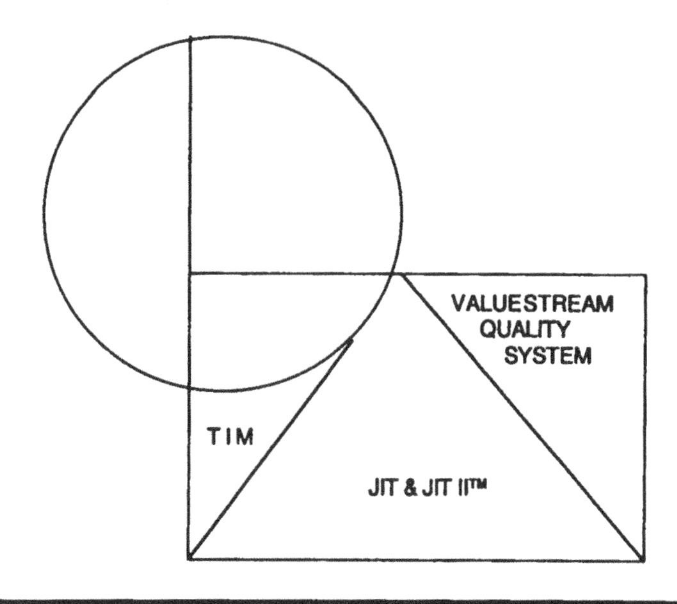

Figure 2.3 Macrologistics Management: Mobilization — Phase II

JIT II™. The technique provides an ongoing dialogue for continuous improvement with suppliers that can deliver hidden competitive advantages to a company. These advantages were trapped by the fact that communication processes did not move across organizational lines and with suppliers due to bureaucracy.

The JIT II™ process captures these opportunities for product improvements cost savings and inventory management. It allows a new culture to form around the JIT II™ inplants. The inplant is an agent provocateur or catalyst for change that in many ways can revitalize the change process. It is therefore an integral technique to developing a Macrologistics System. One of the pioneering applications of JIT II™ is at Bell Labs. Many other companies such as IBM, Intel, Ford, and Honeywell are taking advantage of this technique. Unfortunately, many companies use JIT or JIT II™ as their only Macrologistics strategy. They have mobilized without alignment or integration. And a stool with only one leg is not much use as a chair.

Although the mobilized supplier team is fundamental to many companies, in some cases, such as Whirlpool and Hewlett Packard (Profiles 8.1 and 8.2), the use of internal cross-functional teams has helped generate opportunities for value based strategic initiatives. Customer data and team meetings can be used to innovate across the supply chain. Where the supply chain is shifted from an internal focus on logistics and cost related improvements that are internal to an external customer phased process, it is a value chain. Macrologistics strategies easily emerge from the use of customer definitions of value. These concepts can be used to redefine the corporate goals and they are therefor, useful in delivering change.

Again, the definition of a value-based initiative does not imply that there is alignment and there may not be much acceptance of the resulting initiatives without alignment. Creating mobilized resources such as supplier partners, cross-functional, externally oriented teams, or *Keiretsu*-like new corporate networks, are useful tactics but they do not form the sole basis of strategic change. Examples of these approaches are contained in Chapter 8.

Macrologistics Integration

The most ambitious companies such as Motorola and General Electric are determined to use the latest methods to develop competitive advantage (see Profiles 9.1 and 9.2). When the chairman of GE, John Welch says that he wants his company to be World Class, there is a need to define standards that exceed those of competitors. Once change is desired, the question is what form of change will it take to overcome and stay ahead of competition. Chapter 9 describes how these companies use the Benchmarking technique to obtain data on these incremental changes and how to implement change. Benchmarking is a useful data collection process for the mobilization team and it helps define the goals that will be world class. Competitive advantage is not valuable unless it is significantly ahead of competition. In the case of Motorola, the use of a six-sigma standard for defect reduction is considered World Class because it moved Motorola in to a category of defect-free production that had not been previously attained by an American manufacturer.

Although benchmarking is not a sufficient technique for change, it is necessary to define the kind of breakthrough change that would have a significant competitive advantage. Companies such as General Electric will use this approach to maintain leadership in their industry and to continually distance themselves from competition as shown in Figure 2.4.

It is unlikely that a company can become World Class without the last phase of the change process — Integration. Companies such as Xerox and IBM are now engaged in major integration process using the latest techniques of process reengineering to redefine key process that they use for competitive advantage. Unless they maintain World Class status, lower priced competitors

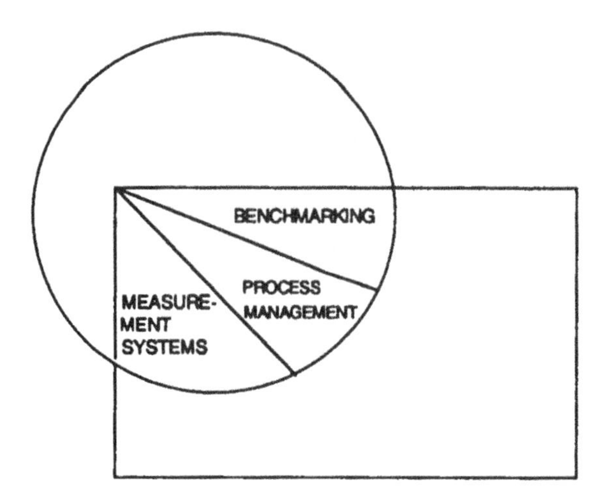

Figure 2.4 Macrologistics Management: Integration — Phase III

are likely to capture large chunks of market share from these major companies. Chapter 10 describes how these companies are now using process management and reengineering to change.

Just as a car needs a tuned-up engine, a full tank of gas (Mobilization), it also needs to have gauges that can measure progress. The design and use of measures such as balanced scorecards, helps facilitate change by creating an environment of measurement of progress. Mars and Air Products use these techniques to guide the implementation of strategy (see Profiles 11.1 and 11.2). If continuous improvement is combined with a measurement system, and World Class goals are defined, the car can proceed at an accelerated pace. Progress toward achievement of goals is hampered by the lack of an evaluation based system. Process efficiency and cost-only scorecards do not work. It is also necessary to evaluate the impact of change on customer satisfaction, quality, and speed.

Summary

Chapter 12 summarizes the process of developing a Macrologistics strategy. In the beginning, companies can have as a strategic goal the use of Macrologistics to gain competitive advantage. This is the case of Federal Express. Then companies can develop a systematic change process using the change model in this book, alignment mobilization, and integration. To develop a comprehensive master plan containing all the necessary elements of this strategy, it is necessary to conduct a diagnostic assessment of the existing initiatives, perhaps using benchmarking to gain information and to formulate World-Class goals.

Mobilization of resources is facilitated using techniques such as JIT II™ and supplier partnering. Integration then becomes the major challenge. As described in this book, the continuous improvement process, Hoshin Planning or (Policy Deployment) and Business Rules Analysis become major techniques for achieving integration. When cross-functional teams are used in an adaptive learning process, new information is continuously obtained on the result of solving problems addressed by the teams. Adaptive Engineering becomes the major diagnostic process for continuously analyzing the results developed by the cross-functional teams. The continuous process also works as solutions are experimented with and the way toward achieving World Class goals is pursued.

Total Innovative Mannagement is a useful approach for designing Macrologistics solutions using participative, cross-functional teams. Management can identify disconnects in alignment, mobilization shortfalls and areas where more integration is needed. The adaptive learning philosophy, when combined with hoshin planning and quality function deployment, enable the integration toward World Class goals becoming a reality. When these goals involve the need to develop a breakthrough and there are macrologistics strategies involved, adaptive engineering and policy deployment can become the engine for change. Whether adaptive engineering is used or not, if breakthroughs in Macrologistics occur, they are likely to provide major competitive advantage. Companies can use these concepts to develop a catalyst for change. The catalyst can be the use of the change model, involving only one initiative or it can be the entire macrologistics management process. The use of all phases coupled with the use of the measurement approaches defined herein will ensure that the World Class goals are achieved, as shown in Figure 2.5.

Lasting corporate change is extremely difficult accomplish. As a manager or executive, you are trying to change the inertia of a moving aircraft carrier armed only with an idea. Hopefully, the use of the Macrologistics framework will make the process of achieving fundamental and significant change possible. We are optimistic that many additional companies will see these benefits and adopt these concepts either partially or wholly. We are interested in developing dialogues with those who are interested in a more comprehensive approach systematic approach to Macrologistics change. Perhaps you will join these pioneering companies and take advantage of one of the most powerful new tools of the 21st century: Macrologistics Management.

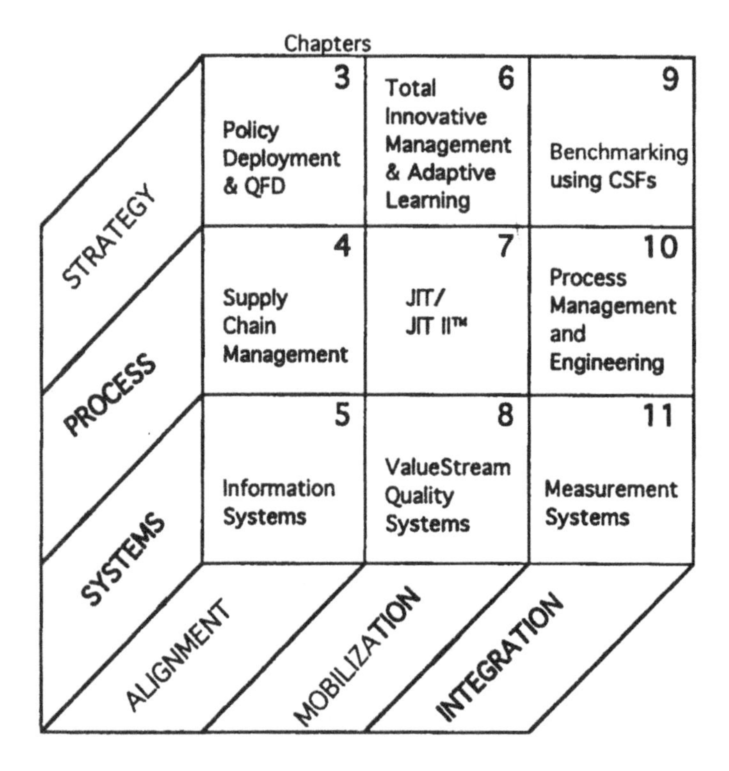

Figure 2.5 Macrologistics Management Alignment Matrix to Achieve World Class Goals Is Shown as a 3-D Cube Illustrating the Interconnectivity of the Nine Major Components

ALIGNMENT

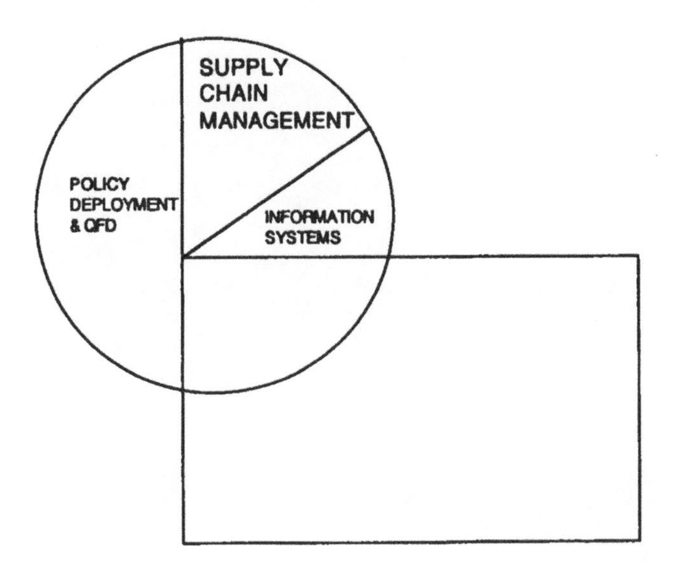
SUPPLY
CHAIN
MANAGEMENT
POLICY
DEPLOYMENT
& QFD
INFORMATION
SYSTEMS

3 Policy Deployment and QFD: The Twin-Engines of Macrologistics

Policy Deployment is a top-down, bottoms-up method to ensure that plans and strategies are successfully implemented within the organization and throughout the value chain. The Japanese deploy their strategy through a process called "*hoshin kanri*" which is also known as "*hoshin planning*," with the word "*hoshin*" referring to policies of management and "*kanri*" to the aspects of planning. Thus, Policy Deployment is a planning and implementing method that ties improvement activities, usually requiring breakthrough results, to the long-term strategies of the organization.

Quality Function Deployment (QFD) is a tool that assists business and service industry personnel in understanding and managing customer requirements and satisfaction. It started with a method developed by the Japanese to assure consistent quality of their automobiles and components, beginning with the initial engineering design. In other words, QFD brings together engineering, design, marketing, sales, and manufacturing people to design and build a product the meets customer requirements every time, all the time.[1]

Description of Policy Deployment

Overall, Policy Deployment requires five basic types of work insofar as Macrologistics is concerned.[2]

1. **Understanding the customer** — this is the logistical focus of strategic planning, analyzing customer needs, competitor's position, and environmental forces.
2. **Goal setting** — reviewing past performances on objectives, critical success factors and targets, while changing work processes to close the gaps between targets and desired performance.
3. **Catchball within the organization** — involves the back and forth deploying new goals and objectives to operating areas and negotiating (catch-ball) with current process capabilities to allow for new levels of achievement.
4. **Monthly and daily management activities** — in order to measure and track how much of the year's objectives are being accomplished.
5. **Checking, inspecting and problem-solving** — the analysis and solving involved in both daily and monthly management.

Policy Deployment starts with the top management team, who is responsible for developing and communicating the vision, as well as building an organization-wide commitment to achievement. The vision is then deployed through the development and execution of annual "*hoshins*", or policy statements. Through the catchball activities, all levels of employees actively participate in generating strategies and action plans for the attainment of the vision and associated objectives.[3] Thus, at each level, progressively more and more detailed action plans are developed to make the strategies and *hoshins* come alive.[3]

Policy Deployment and Goal Setting

Although Policy Deployment starts with the CEO/President in the form of a few critical *hoshins*, each organizational level sets priorities on the areas needing significant improvement or breakthrough. Each *hoshin* contains an objective statement, a goal or target, a strategy, a measure and the person responsible for achievement. As plans are cascaded downward, there should be a clear link the vision and common goals throughout the organization. It is this focus on a few critical goals that achieves the breakthrough desired, by the alignment of critical resources in a focused, concentrated manner.

Goals specify numerically the degree of change that is expected. Thus, the term "stretch goal" is often used to describe the challenging nature of the strategies needed for goal achievement. Measures are used as specific checkpoints to ensure the effectiveness of the individual elements of the strategy.[4] At each successive level of the organization or enterprise, managers develop their own *hoshins*, goals, strategies and measures, and the process is repeated throughout each business unit until the plans are complete for each operation.

Policy Deployment and Alignment

According to various experts on the subject, such as King, Akao, Collins, and Sheridan,[5] the main objective of Policy Deployment is to align the organizational forces of the extended enterprise in the same direction so that everyone within the enterprise knows and understands the overall direction of the organizations involved and how each can play its part to best support that direction.[6] This includes the selection of suppliers who share the vision of the enterprise and are willing to work together to reduce cycle time and prioritize its improvement efforts to tie to the vision. The two factors which can influence this prioritization are importance and the need to improve.

Importance is governed by, among other things, the tie to the vision and major business processes, comparisons by customer groups, critical success factors, and benchmark positions. The need to improve is governed by the levels of customer satisfaction, state of statistical process control, process capability, trends and the overall business environment.[7] Each of the major processes within the value chain needs to be prioritized in order to focus on the high leverage critical success factors that may exist. The use of Critical Success Factors (CSFs) can greatly assist with the prioritization and focusing activities and should be viewed as the strategic, enabling arm of Policy Deployment. See Chapter 9 for more detail about the use of Critical Success Factors.

The missing ingredient in many situations is the inability to correctly understand the customer needs and wants, and to listen to the voice of the customer. In other words, a company should have had a better system for surfacing and aligning their CSFs. Such an approach is found with Quality Function Deployment (QFD) which can be effectively used to surface CSFs and develop Breakthrough strategy while creating policy alignment.

Overview of QFD

QFD is a both a system and a method. As a system it is used for translating customer requirements into appropriate company requirements at each stage of the production-development cycle, from R&D to engineering, manufacturing, marketing, sales and distribution. It is also a planning method or group of techniques for planning and communicating that combines resources and facilitates the focusing of corporate energies in order to better make and deliver products and services, with a high degree of acceptance, at the lowest possible cost. By making customer needs and requirements the prime requirement, satisfaction is more easily attained and a more lasting effect is achieved. It is also a fully integrated technique which consistently correlates measures of performance with predetermined customer needs and wants.

Like Policy Deployment, it utilizes the team approach, relying heavily on brainstorming, Pareto analysis, cause-and-effect principles, and consensus.[8] These concepts determine both internal and external measures of customer satisfaction. The QFD team discusses and analyzes products or services, detailing customer requirements, needs, expectations, and perceptions. Internal measures refer to process-oriented, systematic (operational) elements in contrast to external measures that are derived from customers through surveys, focus groups, and intensive competitive/comparative analyses. Customer feedback provides the control link to company performance. (For additional information, see profile on QFD at the end of this chapter.)

Traditionally, QFD involves an evolutionary process of defining four key phases: design, planning, process, and product, with increasingly sophisticated methodological complexity. Each phase represents a further refinement of key characteristics (variables) that are critical to the customer. The QFD team refines these determinants through piggybacking, synthesis, and expansion. Broad product-development objectives are broken down into specific, actionable assignments via a comprehensive team effort, and without the team approach, it loses much of its power. The process is accomplished through a series of matrices and charts that deploy customer requirements and related technical requirements all the way from product planning and product design through the shop floor, and ultimately out to the distribution channels.

A simplistic application of the QFD principle serves the purpose of illustrating its use. Systematically transferring customer requirements into tangible measures, that can be related to specific performance within the R&D department, can assist in improving customer satisfaction. By listening to the customer in order to understand and interpret the customer's needs and wants, the principles of QFD can benefit both the customer and the R&D organization.

For this application, three key components detail the QFD process. The first element is an intensive focus on the entire dimension of customer satisfaction. This phase involved data collection and synthesis, competitive awareness, customer feedback, and frequent internal/external dialogue. The second phase exists concurrently with the first phase and involves the use of multifunctional teams (including customers). The third phase (ongoing) encompasses detailed advance planning to include follow up, frequent evaluation, customer participation, and review.

Constructing a QFD into a grid format requires detailed information on two elements:

1. Customer needs, wants, and desires.
2. Business deliverables that match customer requirements and expectations.

The following five activities, similar to those found in Policy Deployment and CSFs, are necessary for achieving the full potential of QFD:

1. Customer understanding
2. Goal setting
3. Catchball
4. Monthly and daily management
5. Checking, inspecting and problem solving.

According to Greene[9] and others, this changes how organizations view QFD. When viewed as the product-development process involving a sequence of matrices, strange things occur. Instead, QFD should be seen as the "percolation of customer requirements across any functions of the business, horizontally. If not, QFD becomes a series of matrices turned into a database, usually computerized, thus short-circuiting the customer contact, worker education, and catch ball at its core."

After assembling the initial information, the QFD process begins by listing customer requirements. Requirements originate from the customer's expectations and perceptions of product and service performance. These requirements form the basis of what will satisfy the customer. Customer requirements should be formulated by asking three simple questions:

1. What do customers really want?
2. Why do they need it?
3. How do they know when they have received it?

Requirements define needs, wants, and desires instead of simply solutions. A defined and measurable benefit must be associated with each customer requirement. Customer requirements include basic needs and wants (not generally verbalized) and those additional requirements (often verbalized) that truly satisfy. In addition, requirements that surprise or excite (delight) the customer generally yield a competitive edge. Although focus groups of customers and surveys can provide this information, it is critical that researchers place themselves "in the shoes" of their customers to verbalize these requirements.

Concurrently, the QFD process lists those business deliverables that will be measured to gauge customer satisfaction, productivity, performance, and efficiency. Business deliverables represent internal measures of R&D performance. By assigning a priority to these, QFD evaluates for effectiveness, accuracy, and substance.

Customer requirements, needs, and expectation form the vertical axis of the QFD, with business deliverables listed horizontally. Each customer attribute must correspond to one or more business deliverables along the QFD grid. This phase represents the most dynamic and complex step in the QFD process.

While QFD is not considered a high technology, it has a place in the high-tech arena due to its focus on information transfer as being critical to high-technology processes even more so than with traditional processes. Thus, it can often create technology investments in Information Systems and Technology that add value to the products and services throughout the entire supply chain. Don Clausing, who brought QFD to the U.S. in 1984 following a visit to Japan with expert Dr. Hajime Makabe, cautions on the overuse of technology as a panacea: "We have become good at what I call medium tech. We tend to want to retrench to high technology because we feel its where we still have an edge."[10]

The major difficulty in using the QFD technique results from the highly evolutionary, interactive process of choosing customer requirements and business deliverables. Choosing customer requirements involves both the customer (end user) and the researcher, management, or other suppliers. The process of choosing, evaluating, and finalizing choices is a distinctive learning process. It is not atypical to find that organizations offer solutions to customers prior to determining their true needs. This results in a distorted and biased view of customer needs. By "fitting" solutions to unwanted needs, the customer is distanced from the QFD process and left unsatisfied. The cognitive and emotional aspects of customer satisfaction require a thorough review of potential customer needs prior to the QFD process.

Critics also contend that the Taguchi experiments do not always work in the refinement stage of experimentation. To align the last few minor refinements, a full-factorial experiment, instead of Taguchi's partial-factorial, is necessary. The authors suggest multiple regression, which is, in some situations, almost equivalent to a full-factorial method.[11] For these reasons and others, the Taguchi experiments are often combined with traditional QFD, especially in the design of experiments phase.[12]

Detailing customer requirements and business deliverables assists the QFD process by providing measurable information to track overall performance. Those key characteristics with strong relationships form the nucleus of a complete customer satisfaction program.

In traditional applications, QFD produces a graphic representation of its endeavors. The QFD process yields what is commonly called the House of Quality. The House of Quality aligns customer needs, business deliverables, competitive pressures, and product/process targets listed, with the importance of each characteristic indicated. A traditional House of Quality graphic representation of the QFD process is displayed in Figure 3.1.

The benefits of using such a tool include:

1. Improved lead time (concept to market).
2. Flexibility.
3. Accurate, reliable measures of provider and receiver performance.
4. Competitive analysis.
5. Critical evaluation of business operations.
6. A method for measuring customer satisfaction and product/service performance.

The QFD process involves synthesis, data gathering, and communication. The key concept exposed by QFD is the development and alignment of a communication network between supplier and customer. QFD promises success if customer and performance both interact and interrelate. Measures of performance, aligned with QFD and the customer, provide a road map (benchmark) for success.[13]

Taguchi Methods and QFD

Quality Function Deployment is often used in conjunction with Taguchi methods in Japan, and to some extent in Europe and the U.S. as well (especially

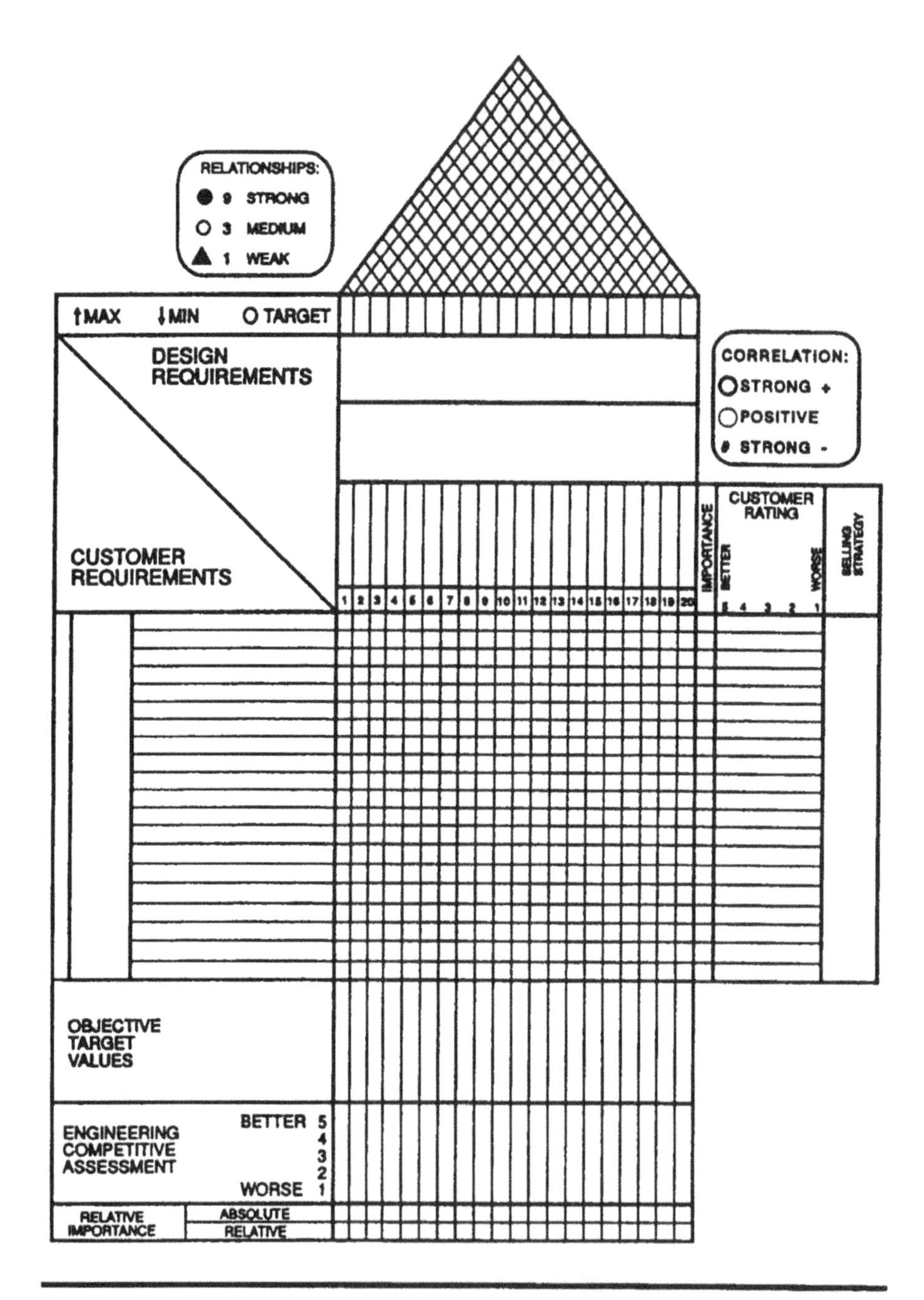

Figure 3.1 House of Quality Requirements Matrix

in the more advanced logistics companies) in that the two processes are complementary in nature. While QFD identifies the relationships between inputs (the hows) and outputs (the whats), it also identifies the conflicting inputs that must be balanced. Therefore, many issues are raised during the QFD process such as:

- What is the nature of these relationships?
- What is the best value of the how items?

Taguchi methods are used to define the nature of those relationships and optimize conflicting inputs by desensitizing the outputs of uncontrolled inputs, thus helping to reduce performance variation. The overall resultant is reduced costs and improved performance and quality.[14] The overall impact upon the value chain can be enormous, as practicing organizations have sometimes found:[15]

- ITT has trained 1,200 engineers in Taguchi Methods and completed 2,000 case studies, resulting in a savings of $35,000,000, according to an ASI study in 1988. Since then, it is estimated that another $35,000,000 has been saved. This amounted to about $1,700 per case.
- **Flex Technologies** has trained 12 engineers in Taguchi Methods and has completed 75 case studies resulting in a savings of $1,400,000, or about $2,000 per case.
- **Sheller-Globe** has trained 120 engineers in Taguchi Methods, completing 225 case studies for a savings of $10,000,000, or about $4,000 per case.

Dr. Taguchi's methodology was first introduced in the U.S. in 1981 after practicing in Japan for a number of years (his work has earned him 4 Deming Prizes among other honors). He was embraced by the auto industry, who at the time was being heavily influenced by the teachings of Dr. Edwards Deming as well. The American Supplier Institute (ASI) became an early advocate of his methods and hosted the First Symposium on Taguchi Methods in 1983, at which the American Taguchi Case Studies were first presented. Case-study preparation and presentation, and the resulting technical interaction, were important aspects of his "learning organization and innovation engineering" philosophy.

His methods are a combination of statistical methods and innovation engineering principles that, when properly integrated, achieve rapid improvements in cost and quality by optimizing product design and operating practices throughout the value chain. Thus, Taguchi Methods are a combination of a philosophy and a collection of tools used to carry out that philosophy as reflected in the following principles:

1. We cannot reduce cost without affecting quality.
2. We can improve quality without increasing cost.
3. We can reduce cost by improving quality.
4. We can reduce cost by reducing variation; and when we do so, performance and quality will automatically improve.

Taguchi disagrees with the traditional "conformance to specification" philosophy advanced by Juran and others. Instead, his approach is to strive for minimal variation around target values without adding costs. The Taguchi Method advocates define quality as "loss to society" from the time a product is shipped. The loss to the customer results whenever a product fails to perform exactly as promised, no matter how small or large, and results in replacement, repair, or an environmental hazard. The loss to the company exists whenever a product or service is overdesigned or is inferior, heavier, less efficient, or larger than necessary.

All loss is eventually experienced by the company through warranty costs, customer complaints, litigation, loss of market, and eventual loss of reputation and market share. Quality is best when product characteristics deviate from target values, quality decreases, and customer dissatisfaction and loss increase. One of the main tools used is the Quality Loss Function (QLF), which is an enhanced cost-control system to quantitatively evaluate quality by assessing the quality loss due to deviation of a quality characteristic from its target value, and expressing the loss in dollars and cents. It is designed to quantify annual cost savings as quality characteristics improve toward target values, making it a suitable tool for evaluating quality at the earliest possible stage of product/process development. Thus, reducing sensitivity to variation is one of the major thrusts of the Taguchi Method.

Using these methods, sensitivity to variation is reduced by adjusting factors that can be controlled in a way that minimizes the effects of factors that can't be controlled, resulting in "robust design." Factors that can be controlled are called "control factors" while those that are difficult or impossible to control are called "noise factors". These noise factors generally cause product characteristics to deviate in **alignment** from target values, which causes variation and quality loss in the following manner:

- **External** — such as temperature, humidity, operator error, vibration, etc.
- **Internal** — such as wear, deterioration, etc.
- **Product-to-product** — due to part-to-part vibration

Control factors which reduce or minimize the effects of noise factors and thereby reduce variation are selected during "parameter design" with the intention of finding the best combination of materials, processes, and specifications that produces the most stable and reliable performance throughout the value chain at the lowest cost through experimentation. This is generally accomplished by maximizing a measure called the "signal-to-noise ratio" (S/N Ratio), which is derived from communications technology and describes the signal as the desired output and the noise as whatever gets in the way. Thus, the S/N Ratio measures the stability of a quality characteristic's performance and the QLF is used to calculate the effect of that stability in monetary units. High performance is measured by a high S/N Ratio and means low loss, concluding that the larger the ratio, the more robust the product against "noise." Tolerance design methods are used to test for more costly materials with tighter tolerances.

Summary

As mentioned earlier in the chapter, the model of the five fundamental work force tasks **realigns** not only how QFD is viewed, but also how Policy Deployment and logistics are viewed as well. It is not merely a cascade of objectives but a unification approach between strategy, process and logistics both horizontally and vertically. Therefore, it contributes to the likelihood that alignment occurs on key operating principles. If Policy Deployment and QFD are added to older systems of process support, they add complexity and slow things down while focusing and **aligning**. However, if they are replacements for older systems of logistics management, they can radically simplify the documentation and performance of work. However, because QFD does not do more than a superficial level of alignment, Hoshin Planning and other techniques are also required.

All too often, Policy Deployment gets distorted into a complex of committees and policy initiatives that the workers are told to implement. This creates unneeded bureaucratic structures that confuses the workforce and slows down implementation to a crawl. The four elements of Macrologistics Management just described can be aligned into one comprehensive framework going in four directions. Taken together, they provide four times the focus and six to eight times the productive output for about one-third the effort and workforce learning.[16] The alignment process is also continuous, which suggests that an innovation-based adaptive learning approach is also needed, as described in a later chapter.

References and Endnotes

1. *Total Quality in Research and Development*, by Gregory McLaughlin, St. Lucie Press, 1995, pp. 171–172.
2. *Global Quality*, by Richard Tabor Greene, ASQC Quality Press, 1993, pp. 831-832.
3. *The Management and Control of Quality*, by James Evans and William Lindsay, West Publishing, 1993 (second edition), pp. 166.
4. Ibid. pp. 187.
5. These three authors represent divergent yet complementary viewpoints concerning the implementation strategies and ultimate value of Policy Deployment. In *Hoshin Planning: The Developmental Approach*, GOAL/OPC, 1989, Bob King discusses the origins and Americanization of 'hoshin' planning. Yoji Akao, in his masterwork *Hoshin Kanri-Policy Deployment for Successful TQM*, Productivity Press, 1992, approaches the subject from its developmental origins in the practices of Japanese administrative business management. Finally, both Bruce Sheridan and Brendan Collins (in related writings) offer a practical working model designed for western management, which is based upon the FPL Policy Deployment process, in *Policy Deployment: The TQM Approach to Long-Range Planning*.
6. *Total Quality in Purchasing/Supplier Quality Management*, by Ric Fernandez, St. Lucie Press, 1995, pp. 97 and 100.
7. Ibid. pp. 100–102.
8. The point being made here is that QFD requires both the soft and the hard quality tools, often in combinations at the same time. Brainstorming is used heavily to generate the flow of ideas and information. Paereto analysis is used to align and stratify the data into possible groupings. Cause and effect analysis is used to perform two and three dimensional problem solving. And consensus is used to align the data with the people's beliefs and biases.
9. This premise and accompanying analysis is very clearly brought together by Richard Tabor Greene in *'Global Quality'*, pp. 832-839. The major premise is unification, that is to use the five basic tasks of PD and QFD to eliminate dissipation of those methods into complex procedures and committee work. This concept was developed in the Japanese manufacturing industry during the 1980s

and at Ford pioneered it in this country about ten years ago. These tasks can be used as one process going in two directions. The ultimate goal is to unify and align the 'four deployments' of QFD, Policy, Automation Deployment and Technology Deployment as one process in four directions.

10. The Customer-Driven Company, by Eureka and Ryan, pp. 7.
11. For further details, see "What Hath Taguchi Wrought?", by Eugene Sprow, *Manufacturing Engineering*, April, 1992, pp. 57–60. What is often under attack, concludes the author, is not Taguchi's ideas but their glamorization by entrepreneurs.
12. See *Taguchi Methods: A Hands-On Approach*, a book by Glen Stuart Pearce. The first chapter provides a conceptual framework for Taguchi's contributions to quality. The remainder of the books 23 chapters are divided into four sections which align with the steps involved in designing, conducting, and analyzing experiments designed to isolate problems and optimize the manufacturing process. This book contains three large appendices on orthogonal arrays, linear graphs and interactions between two-column tables.
13. *Total Quality in Research and Development*, by Gregory McLaughlin, St. Lucie Press, 1995, pp. 172-176.
14. *The Customer-Driven Company*, by William Eureka and Nancy Ryan, American Supplier Institute, 1988, 1992, pp. 103–106.
15. Many American and European companies are implementing Taguchi Methods, however, because they contain proprietary information, the results of these applications cannot always be made public. It has been estimated that between 5,000–10,000 case studies have been completed in the U.S. alone, although the number of case studies prepared annually in Japan has been estimated to be 20 times higher.
16. *Global Quality*, by Richard Tabor Greene, pp. 835.

FPL: The Light That Flickered* 3.1

In 1989, above the office doors and on every desk and remote location was the twenty-word challenge that was to change the course of quality history in the U.S. and beyond: "**We will be the preferred provider of safe, reliable, and cost-effective products and services that satisfy the electricity-related needs of all customer segments.**" This was a vision for the ages but would not even last a decade. The Quality Policy and system that won the Deming Prize and the Edison Award turned on its masters, became a bureaucracy, and became the first major downsizing effort of the 1990s.

FPL used its logistics systems to effectively integrate in a vertical manner its generation, transmission and sale of electricity for use by individual customers, businesses, and industries. State of art Information Systems, a highly

* ©Stategy Associates.

automated physical distribution system, coupled with a well integrated Policy Deployment/QFD system, became the cornerstone for advancing FPL's successful bid for the Deming Prize in 1989. Macrologistics also played a key role in the completion of its nuclear power plants in the 80s, as well as its downsizing efforts of the 1990s.

The well-integrated Information Management system, in which FPL invested almost $100,000,000 during the Deming Prize era, helps provide critical data for the regulatory agencies and its Corporate Measurement System provide for both strategic and daily control. Cross-functional management and Breakthrough team work helped the utility save hundreds of millions for its customers, while improving service and reducing cycle time on key activities.

Due to Florida's sub-tropical climate, FPL experiences heavy demand loads during the months of June through September, when air-conditioning usage is high. The company's annual system peak also occurrs for very short periods during the winter months on cold days when customers use electric heating. This saturation of air-conditioning, electric heat, and other appliances results in FPL having one of the largest rates of electricity consumption per residental customer, thus affecting the logistics of the load planning process.

During the early 1990s, the FPL Quality Improvement system received a blackeye across corporate America when Chairman Jim Broadhead began the downsizing cycle, but the company has successfully rebounded, both in public image and stock price.

Brief Corporate Description

FPL Group is a holding company, the principal subsidiary of which is Florida Power and Light Company (FPL), one of the largest investor-owned utilities in the nation and the largest utility in the state of Florida. It is the fastest growing electric utility in the Nation in terms of customer accounts. Also, the company has the highest annual turnover rate of customers. This growth and turnover rate affects construction activity, and the number of service personnel and logistics support required to provide service.

The FPL service territory covers approximately 28,000 square miles, which is half of Florida, and reaches some 6 million people. FPL operates 13 plants, 400 substations, and over 50,000 miles of overhead and underground transmission and distribution lines. Annual revenues are almost $6 billion and the total plant in-service worth is approximately $12 billion.

Profile in Detail

An electric utility has the legal obligation to serve all present and future customers by providing electricity at any time and in any amount required within its designated service territory. The obligation to serve and the need to generate and deliver electric power at the same instant the power is demanded is a distinguishing feature of utilities and provides a perfect application for Macrologistic Management principles. Logistics management helped FPL maintain very high standards of reliability in their generation and delivery systems to consistently serve the demand of its customers. Additionaly, to meet customers' future needs and requirements, utilities must develop long-term supply plans since it takes many years of logistics preparation to produce additional sources of generation capacity.

In the U.S., electric utilities are highly regulated. FPL's retail rates are regulated by the Florida Public Service Commission (FPSC) and its wholesale rates by the Federal Energy Regulatory Commission (FERC). Nuclear operations are regulated by the Nuclear Regulatory Commission (NRC) and enviornmental areas are regulated by local, state, and federal regulatory bodies. FPL's Macrologistics Management strategy plays a key role in dealing with all this regulation in a consistent manner.

Business Situation Leading up to QIP Macrologistics

In the period from 1946 to 1974, FPL experienced rapid growth of double-digit proportions, making it difficult to keep up with the need to plan, finance, construct, and operate an electric system in a growing South Florida community. Throughout this period of rapid expansion, FPL maintained stable prices for its customers until the oil crisis of 1974 came along with its period of high inflation. In 1978, the government passed the National Energy Act, which resulted in competition for utilities and promotion of conservation of electricity.

Due to these conditions, the trust of FPL had developed with its customers suffered. During the 1970s, the company was forced to increase utility rates repeatedly because of increasing costs, slower sales growth, and stricter federal and state regulations. FPL had become bureaucratic and inflexible. In 1981, Marshall McDonald, then chairman of the board, realized that the company had been concerned with keeping defects under control rather than improving quality and addressed the problem head on:

> "As FPL grew, we had become more bureaucratic and cumbersome. We were often inflexible in our operations and often treated all of our customers the same, regardless of their individual needs. At the same time, we could forsee no significant technological innovations to reduce the escalating power supply. A change in management philosophy was needed to achieve customer satisfaction and reduce costs through greater management effectiveness.

> We concluded that the company's internal and external environments were changing faster than the organization could adapt and that corporate goals needed to be established and achieved using new management techniques. In order to achieve our new corporate vision, goal setting was introduced to change the corporate culture as part of the introduction of our QIP program."

Due to his concern for quality, MacDonald introduced Quality Improvement teams in 1981, based upon the combined teachings of Deming and Juran. The objective of the Quality Improvement Teams were to provide a structured environment for employees to work together in a new way. MacDonald's vision was to move the organization toward the following four value-added goal-setting areas:

- Improving the quality of products and services
- Developing the skills of employees
- Promotion communication and teamwork both within and between suppliers
- Enhancing the quality of worklife

Management knew this was a step in the right direction, but such goals alone would not bring about the change needed for the company to survive. MacDonald tried to convince other executives that a total quality improvement process was needed, but all the experts that FPL talked to were in manufacturing, while FPL was primarily a service company. In 1983, while in Japan, MacDonald met the president of Kansai Electric Power Company, a Deming Prize winner, who told him about their total quality efforts. Company officials began to visit Kansai regularly, and with their help, FPL began its own company-wide TQC logistics-based effort, known as the Quality Improvement Program (QIP) in 1983. MacDonald emphasized the following changes:

> "FPL had to change our way of thinking from supply-oriented to customer-oriented; from a power generation company to a service company. We needed new logistics strategies to provide a means of addressing the key issues surrounding the satisfaction of customer needs and expectations. We began our renewed quality efforts in 1981 with a limited and narrow approach called Quality Improvement teams, similar to Japan's Quality Circles. We soon realized, however, that QI teams alone would not achieve the results needed to change the company."

Policy Deployment Enters the Scene

"Policy Deployment" was the driving force — or the engine — behind the full development of the QIP program. In 1985, FPL chairman John Hudiburg (who reported to MacDonald) announced that the company would broaden the QIP program to include Policy Deployment:

> "After thoroughly investigating TQC methods, I am convinced that these approaches, coupled with a Policy Deployment approach, are the best management systems ever conceived. The QIP philosophy and related systems would be deployed throughout FPL and would be the method used to accomplish the corporate vision. Our primary motivation in introducing an expanded QIP was to establish a management system and corporate culture to assure customer satisfaction. A fundamental change would be needed to listen to the voice of the customer and to identify their needs and expectations. As a result, a number of Policy Deployment short-term plans are being initiated to address priority issues."

Policy deployment is the cornerstone of FPL's management system. It is a management process for achieving breakthroughs on major corporate problems and focuses on customer needs by deploying resources on a few, high priority issues. It is a method that takes corporate vision and determines priority issues that will make the vision a reality. For FPL, the issues involved improving reliability, customer satisfaction, and employee safety while keeping costs in control. Each department was then responsible for developing plans to help improve these areas. Once plans were determined, their status was checked regularly to make sure they were on schedule. Each department was limited to working on no more than three items that had the most influence on their department's performance, but the work on these was expected to be done in great detail.

At FPL, Policy Deployment targets the achievement of breakthrough by concentrating company resources and efforts on a few priority areas. According to Hudiburg, the early focus was to:

- Increase performance levels
- Improve communication of company and department direction
- Improve coordination within the company and the value chain
- Attain broad participation in the development and attainment of corporate goals

Prior to 1985, the company used Management by Objectives (MBO) as the principle method for achieving corporate objectives. While MBO focused on the company point of view, it did not adequately consider the customers viewpoint. Also, it did not provide a systematic process for insuring that corporate objectives were met. Policy Deployment was introduced to provide a process for achieving corporate objectives and deploying logistics effectively. However, the company attacked too many problems and solved only a few. Also, the early improvement attempts were not focused on customer needs. Therefore, it was necessary for the company to take corrective action for the next cycle and in each successive cycle thereafter. In 1988, cross-functional teams were introduced to help executives coordinate the major activities of FPL from a corporate accountability perspective.

The many improvements that were made clarified how corporate logistics and management activities tied to an overall management system, which was enhanced with each successive Policy Deployment cycle. The following are the four key characteristics that Hudiburg and MacDonald extolled about the FPL management system:

1. **Focus on Customer Needs:** Policies are identified and developed based upon quality elements (QFD), which are established through a combination of the customer's voice and the utility industry's obligation to serve.
2. **Management Reviews:** Improvement activities are reviewed by the office of the chairman, the president, responsible executives and managers to check on the achievement of company policies. We then take action when necessary to promote QIP.*
3. **Cross-functional Management:** Responsibility for improvement objectives are assigned to executives, although activities cross functional lines and boundaries.
4. **Integration of Policy Deployment and Budget:** Through the process of management consensus and "catchball", the resources are allocated to support annual improvement activities.

Logistics Aided by Information Systems Development Help FPL Win The Deming Prize

Although Policy Deployment was implemented to provide a system for customer satisfaction in 1985, it did not provide a system for standardizing the initial improvements and replicating them throughout the value chain, including

* FPL has three levels of Management Reviews. Level I reviews are conducted by the president and executive vice-presidents by reviewing the business plan with emphasis on cross-functional and short-term plans. Level II reviews are conducted by appropriate vice-presidents and are designed for managers to present their progress on the associated business plans. Level III reviews are conducted by department heads with their managers and supervisors to focus on how well QIP has been implemented within their departments, with a particular emphasis on priority problems and the application of daily control systems and process management activities.

the suppliers. This next phase, called quality in daily work (or daily work control) provided a tool for FPL's managers to control their work processes throughout the entire logistics value chain, from customers back through the supply chain.

Daily work control had the following characteristics, as outlined by John Hudiburg:

> "First of all, it focuses on manager's and supervisor's key accountabilities. The idea is to develop control systems for all top priority jobs in order to standardize, replicate and improve all aspects of the daily operations of bringing electricity to our customers. Thus the focus is on the customers needs. The key outcome is also to identify areas for the development of computer systems that free many line employees from repetitive tasks."

FPL's computer systems have provided the necessary tools to resolve some of the difficulties in implementing logistical control systems by giving FPL the ability to:

- Tailor services to meet individual customer needs
- Provide quick access for problem solving at FPL's many remote areas
- Stratify, analyze, and graphically display data
- Replicate processes and standards across multiple-work locations
- Collect, store, retrieve and analyze data consistently throughout the company
- Provide training for repetitive activities
- Maximize the economic logistical operation of the power supply system

Information Systems and Services Organization

In 1989, the year FPL won the Deming Prize, the Information Systems & Services Organization consisted of about 600 employees, of which 288 were in Systems and Programming and 314 were in Computer Operations. A total of nearly $100 million had been spent to bring the entire operating system to a level of peak performance in support of the following Information Systems priority areas:

Systems and Programming consisted of (a) 219 Application Development and Support employees who developed new systems and performed maintenance and enhancements (b) 37 Information Planning employees who ran the Development Center, did DP training and QA, performed user liaison and data security, and helped with planning and administrative services (c) 32 User Access personnel who ran the Information Center, provided PC and LAN support, and offered end-user computing support.

Computer Operations consisted of (a) 70 Computer Center employees who ran the operating systems, the HELP desk and the Network Control Center (b) 58 Technical Systems employees who did systems programming and data base administration activities (c) 147 Data Preparation employees who performed data entry, output control and balancing, and payment processing (d) 39 Telecommunications employees who performed data communication activities, voice communications and fiber optics services.
Operating Environment hardware consisted of (a) three 3090/600 processors — one for MVS batch and development, one for MVS on-line systems and one for VM end-user computing (b) 14,000 terminals and printers and (c) over 3,000 personal computers. The software looked like (a) corporate based MVS, CICS, IMS/DB, DB2, COBOL (b) end-user computing using VM/XA, FOCUS, TELL-A-GRAF, AS, SAS.

Application Programs Used to Support Daily Control Logistics Activities

Application environment consisted of fifty online applications and functionally divided portfolios as follows:

- **Customer and Marketing** — customer billing, customer information system and electronic meter reading
- **Distribution and Construction** — trouble call management, construction management, facilities graphics, and Divisions Maintenance Management
- **Engineering and Generation** — generating equipment management and nuclear control and information management
- **Financial and Personnel** — financial accounting, purchasing, inventory control and accounts payable, as well as human resource management.

As the Deming Prize unfolded, the overall objectives of management, especially in the area of logistics management, was fourfold: change to customer oriented thinking, create resilience to change, seek breakthroughs in achieving cost reduction through greater logistics management efficiency, and to counter bureaucratic attitudes throughout the supply chain.

"Quality in Daily Work" (QIDW) is the expression that FPL eventually used for the concept for improving business systems quality control of daily work processes. As previously mentioned, it involves standardizing and redesigning work routines, removing waste from them, promoting the concept of internal customers, and enabling better practice to be replicated from one location to another. QIDW meant measurement driven control systems consisting of flowcharts, process and quality indicators, procedure standards, and computer systems. By examining and analyzing work over and over again, employees in every area contributed to simplifying their work and improving the value chain.

The goals of the daily control system activities were to apply the Plan-Do-Check-Act (PDCA) philosophy to processes and work activities in order to meet the needs and expectations of customers by:

- Maintaining the gains achieved through improvement projects
- Achieving consistency in operations as well as results
- Clarifying individual contributions toward achieving customer satisfaction
- Improving daily operations throughout the entire value chain

One illustration of how Macrologistics was used was to develop a computer system for processing customer trouble calls. In the system, the computer first checks to find out if the customer has been disconnected for nonpayment, then begins to locate places and devices that may be malfunctioning, and routes the call through a dispatcher to a troubleshooter. A repairman heading to the scene may have a diagnosis before arrival. The information is stored in a database to be used for future improvement planning.

Computer systems were a unique feature of the FPL logistics standardization. Standard worksteps can be automated. Employees all along the value chain can check the results and take the appropriate actions. With the aid of computers, they also analyze data trends. Management uses Information Systems to analyze trends over longer time periods. Through the implementation of Daily Control Systems, the understanding and deployment of customer needs improved significantly with the following results, as reported by John Hudiburg:

> "The customer perspective has been integrated throughout corporate-wide activities. Management is now involved in reviewing the application of daily control systems to control and improve work processes. Also, SPC techniques are applied to stabilize and validate work processes for the purpose of increasing customer satisfaction."

Taking Logistics Control to the Utmost in Nuclear Operations

FPL's nuclear plants use the fission of uranium to produce electricity. The fission products are radioactive and as they decay, they produce high energy radiation. Excessive exposure to this type of radiation can be harmful to the employees and to the public. Logistics management plays a key role in improving the safety factor in this environment. This led to the Prime Nuclear Safety Function. According to John Hudiburg:

"The ability of our nuclear plants at Turkey Point and St. Lucie to produce power depends upon the strictest conformance to NRC regulations. If FPL or the NRC identifies a condition considered adverse to safety, the unit will be shut down until the condition is remedied. The NRC may suspend or revoke our license to operate, as they have five other utility units at the present time."

Suggestion System for Improved Logistics Methods

To help clarify individual contributions throughout the entire value chain, FPL revamped a centralized suggestion system it had been using for many years. Only about 600 suggestions had been submitted annually and it usually took six months for evaluation. A new decentralized system was proposed with simplified procedures to improve the response time. Employees would participate in the implementation of their own suggestions. In 1987, about 1,000 suggestions were submitted; by the end of 1989 this increased to 25,000 suggestions and many of them involved day-to-day logistics-type activities. The results were dramatic. Communication and teamwork improved and the bargaining-unit team participation and involvement reached an all-time high of 43%, up from 10% in 1983. Employees also improved their problem-solving skills and problems which had previously been handled by management were now being addressed by first-line employees, using data instead of intuition.

Training, Training and More Training

Training played an important role in FPL's quality transformation. They found that training enhanced enthusiasm and participation. Supervisors were expected to train their employees and play a more active role as coaches and cheerleaders. As line employees become more skilled in diagnosing and solving problems, issues that once required management attention are now handled by line employees. Problems are dealt with on a factual basis, not with intuition. All employees developed a much broader view of the company and more flexibility in dealing with customers.

The management system also changed. Customer satisfaction became the focus of attention rather than cost control. Management reviews checked on improvement progress monthly. Goals became long term, and progress checks were frequent. Managers reviewed progress with better statistical insight, recognizing that variation will exist, but seeking to rid the system of common causes. Cross-functional teams were used to carry out large-scale improvement projects. Finally, the budget was integrated with quality improvement.

During the past seven years, FPL has maintained its Policy Deployment focus on the following IS related logistics activities:

- Improved availability of customer and supplier information
- Improved availability of Trouble Call Management system
- Improved response time of on-line systems at Trouble Call offices
- Reducing on-line systems communications network unavailability, including EDI links
- Implementing a structured system development projects for new computer systems development, using QFD and Policy Deployment tools
- Improving information center customer satisfaction by reducing the time to resolve customer and vendor calls
- Improving response time of on-line systems at regional phone centers
- Improving inventory turns each year
- Reducing the number of days for unplanned outages at nuclear plants through improved logistics management.

The influence of Macrologistics Management at FPL during the Deming Prize years was considerable. The average length of customer service interruptions dropped from about 75 minutes in 1983 to about 47 minutes in 1989; the percentage of on-line systems network unavailability was reduced by 75% from a high of 0.7% in January of 1985 to a low of 0.1% in 1990 and thereafter; prime-time application outage frequency was reduced to an average of 0.1%; the percentage of customer call-backs not returned within one hour dropped from 70% to 29% and has remained on target; the number of complaints per 1,000 customers fell to one-third of the 1983 level; safety improved; and the price of electricity has stabilized. And at the core of these improvements is the Daily Management System which is founded upon Statistical Process Controls Macrologistics methods.

Summary

The pitfalls of the FPL QIP program were significant and eventually led to its fall from grace, although for the most part the logistics areas have remained unchanged from the Deming Prize years. The main drawback was bureaucracy in that every management system, left to its own devices, tends toward complexity. FPL was no exception. QI was viewed in opposition to doing the "real work". Often the training was incomplete and managers were left out. The staff groups were slow to accept the rigor and discipline and complained that it slowed them down. Also, in many cases, teams selected problems that were too big and tried to "solve world hunger". And Process Management (QIDW) became difficult to apply, except in repetitive areas, until the tools of QFD and Taguchi Methods/Design of Experiments were brought in.

In spite of its shortcomings, the impact upon the Information Systems areas and Logistics functions was enormous. For the first time, customer needs were effectively translated into priority problems that were prioritized and solved, once and for all. Every department measured its impact upon corporate goals and a new focus resulted in customer satisfaction. Results were consistently measured, and process and quality measurements have been developed for most jobs, resulting in a workers sense of obligation to improve the workplace. Management experienced and enhanced awareness of the impact of each group in the value chain, as well as finding itself involved in a lower level, intuition-based, gut-feel decision-making has been replaced by a renewed focus on process improvement, standardization, and use of data. And everyone is rewarded for his or her "bright ideas."

© Stategy Associates. *Sources:* FPL Visitor Orientation, 1991; Brad Stratton, "A Beacon for the World", *Quality Progress*, May 1992; Al Henderson & target stuff, "For FPL After the Deming Prize: The Music Builds," *Target*, summer 1990, pp. 10-21; Neil DeCarlo and Kent Sterett, "History of the Malcolm Baldrige Award," *Quality Progress*, March 1990, pp. 21–27.

3M: The Masters of Innovation — 3.2

While a number of firms, including Hitachi, Merck, Toyota, Banc One, Sony, AT&T, Milliken, Kodak, Barnett Banks, Hewlett-Packard, Intel, Johnson & Johnson, and Rubbermaid, have been highlighted in this book as being very creative, 3M has been mentioned more than any other firm. Indeed, 3M could have been used as the example for each of the forty-nine characteristics of innovative organizations. Most authorities consider it the most innovative firm in the United States, not only in product innovation — it has over 60,000 items in its product lines — but also in process, marketing, and management innovation. Why is 3M so successful? Because it has paid attention to innovation and has created a culture that fosters it. Following is a list of examples of actions taken by 3M to promote innovation as related to each question of the questionnaire. There is a mixture of types of innovation but most are product and process related. Many more could be cited.

1. Has a stated and working strategy of product and process innovation.
2. Constantly spins out new business units with the innovator in a key position.
3. Ties salaries and promotions to innovation. It even rewards other companies for innovation. For example, in conjunction with The Healthcare Forum, it presents an annual innovation award to the most innovative healthcare organization.

4. Focuses the entire company on innovation.
5. Has people who generate ideas. It seeks inquisitive, creative people as employees.
6. Celebrates extremely creative people and their accomplishments, for example, through the Carleton Society and the Golden Step Award.
7. Proactively creates new opportunities and responds to change.
8. Encourages employees to spend part of each day figuring out ways to improve products from the customer's perspective. Researchers, marketers, and managers routinely visit customers, and customers often sit in on brainstorming and other product development sessions. Both TQM and innovation programs start with a focus on the customer.
9. Uses cross-functional new-product teams and process-redesign teams. For example, cross-functional "action" (speed) teams were used to develop a new respirator in record time.
10. Celebrates creative successes through various media, including promotional videotapes.
11. Allows people to make mistakes; it looks for a good batting average rather than a home run every time at bat.
12. Has a large staff devoted to R&D. Has encouraged everyone to be an idea person.
13. Encourages risk taking through various programs, including financial investment in new ideas (project Genesis).
14. Continually creates new products or services and/or enhances old ones. Its corporate objectives state that 30% of its sales are to come from products that did not exist four years previously. 3M has innovation requirements for all product divisions and for process as well as product innovation. Virtually everyone at 3M is involved in innovation and is given the time and direction to be innovative. Products and services are marketed innovatively. For example, 3M sales personnel use floppy disk sales materials, which replace bulky paper catalogs. Technical information is delivered in the same way. Innovative management is practiced. For example, TQM and innovation are treated as partners in creating customer satisfaction. 3M has learned that these two forces work well together to create a competitive advantage.
15. Has objectives for its managers for product and process innovation. One of the firm's major objectives is that 30% of sales must come from products that did not exist four years previously.
16. Has three different layers of product research centers.
17. Has information management information systems that scan the environment for new opportunities, track competitors' actions, conduct benchmarking analyses, keep abreast of new technologies, and exchange information internally. For example, when Dow Chemical chose not to market a unique nonstick coating, 3M licensed the concept from Dow.

18. Suspends judgment on new ideas through a formal process in which new ideas are presented to various groups at various levels of the organization.
19. Uses idea/innovation champions to move ideas through formal processes. For example, Art Fry championed Post-It Notes.
20. Does all it can to separate politics and other factors from the evaluation of ideas. Ideas are reviewed on their merit by several different groups.
21. Seeks continuous, incremental process innovation, but also goes for the big bang in new-product development.
22. Puts every product on trial for its life every five years (or sooner).
23. Stresses maintaining open communication, including positive forms of conflict.
24. Has a formal idea-assessment system for product, process, marketing, and management innovation that, among other things, separates creation from evaluation.
25. Delegates large amounts of authority to division managers, who in turn delegate to their subordinates. Objectives are clearly stated, but how to achieve them is up to the managers and/or subordinates. There are relatively few company policies and little bureaucracy.
26. Trains its people in the use of creativity techniques to generate new ideas.
27. Believes in a creativity ethic: that it has always won or lost through innovation and that unfettered creativity pays off in the end.
28. Engages in knowledge management — identifying knowledge assets, sharing information, and tapping the innate knowledge of individuals. Actively uses information-sharing systems and technology to search for knowledge both outside and inside the firm. 3M encourages its employees to look everywhere for ideas — including competitors, customers, suppliers, and scientific sources outside the company.
29. Makes commercialization or utilization of new processes a priority.
30. Keeps business units small (usually under $200 million in sales), giving them flexibility and the authority to pursue any opportunities they see.
31. Sets financial hurdles for new products, but as only one of several criteria. 3M looks past the numbers to other issues, such as market share.
32. Has a very trusting management style in which employees are encouraged to self-manage and solve problems.
33. Employees use various creative problem-solving techniques.
34. Began managing its culture to make it creative and innovative more than seventy years ago, long before the concept of managing organizational culture became popular. (3M has been in business for over 100 years.)
35. Practices organizational learning through knowledge-sharing sessions.
36. Uses speed strategies and/or almost unobtainable objectives for product, process, marketing, and management innovation.

37. Uses alliances to obtain product, process, marketing, and management innovations; for example, 3M has an alliance with two leading companies of the Sumitomo Group in Japan.
38. Has formal information-distribution systems that require sharing of knowledge among divisions, and encourages informal exchange networks.
39. Uses transformational leadership.
40. Allows selected employees to spend 15% of their time every week for reflection on innovative new products or processes.
41. Insists on constant change if it improves the firm.
42. Leverages resources to achieve seemingly unobtainable objectives.
43. Knows when to lead the customer to new products or services, reduce costs through improved processes, take advantage of new-product opportunities through marketing, and reduce costs through innovative management.
44. Has an effective and efficient structure for evaluating product design, process improvement, and marketing or management innovation.
45. Has effective employee suggestion programs.
46. Manages innovative personnel with special approaches, including a hands-off management style, dual career ladders, a penchant for innovation, and removal of political barriers.
47. Provides physical facilities that are conducive to creative thinking and the exchange of ideas.
48. Requires nonmanagerial employees to have stated objectives for product, process, marketing, and/or management innovation.
49. Invests heavily and appropriately in R&D. (In 1992 the company spent $1 billion on R&D, an amount equivalent to 7.3% of sales.)
50. Uses Macrologistics Management to optimize the value chain.

© *James M. Higgins.* *Sources*:Kevin Kelly, "The Drought is Over at 3M," *Business Week* (November 7, 1994), pp. 140–141; James J. Thompson, "Quality and Innovation at 3M: A Partnership for Customer Satisfaction," *Tapping the Network Journal* (Winter 1993–1994), pp. 2–5; Chris Rauber, "21st Century Vision," *Healthcare Forum* (January-February 1994, pp. 75–78; "Computerized Sales Tools Give Instant Information," *Adhesives Age* (April 1994), pp. 37–38; Philip E. Ross, "Teflon Deja Vu?" *Forbes* (April 11, 1994), p. 130; George M. Allen, "Succeeding in Japan," *Vital Speeches of the Day* (May 1, 1994), pp. 429–432; Gregory E. David, "Minnesota Mining & Manufacturing," *Financial World* (September 28, 1993), p. 58; Michael K. Allio, "3M's Sophisticated Formula for Teamwork," *Planning Review* (November-December 1993), pp. 19–21; "3M Backgrounds: 30 Percent Challenge," internal corporate document, 1993; Steve Blount, "Test Marketing: It's Just a Matter of Time," *Sales and Marketing Management* (March 1992), pp. 32–43; Tom Eskstein, "Reader's Report — 3M's Creativity Takes a Lot of Practice," *Business Week* (May 15, 1989), p. 6; Russell Mitchell, "The Masters of Innovation," *Business Week* (April 10, 1989), pp. 58–64; Lewis W. Lehr, "A Hunger for the New," *Success* (September 1988), p. 12; Alicia Johnson, "3M Organized to Innovate," *Management Review* (July 1986), pp.

38–39; Kevin Kelly, "3M Run Scared? Forget About It," *Business Week* (Industrial/Technology Edition, September 16, 1991), pp. 59, 62; Ray Kubinski, Sam Bookhart, Anita Callahan, Marvin L. Isles, Charles Porter, Anthony T. Liotti, and Margie Tomczak, "Managers Forum Focuses on Competitive Strategies and Continuous Improvement," *Industrial Engineering* (February 1992), pp. 30–32.

4 Supply Chain Management*

Ralph Lewis and Frank Voehl

An important element of Macrologistics is the ability to map the entire supply chain at the systems level. The process by which the supply chain is analyzed and designed is fundamental to the potential ability of the organization to identify opportunities and vulnerabilities that can impact its future. The extent to which opportunities are identified provides substantial leverage capability to the organization in meeting and exceeding competition. Supply Chain Management (SCM), therefore, is an important component of Macrologistics strategy because it provides a blueprint for documenting existing problems and opportunities. The SCM process also helps to create new organizational units that are critically needed to act in motivating policy choices. In this chapter we provide an overview of SCM and illustrations of the successful implementation of SCM.

What Is It?

Briefly, SCM is the systematic effort to provide integrated management to the supply chain to meet customer needs and expectations from the suppliers of raw materials through manufacturing to end customers. It is characterized

* This chapter is adapted from a book in progress titled *ValueStream Management: The New Value Proposition* by Ralph Lewis, Bill Meyer, Jeffrey Vengrow, and Frank Voehl.

by, "concern for the movement and storage of goods, with the associated information flows from the beginning to the end of the supply chain."[1]

Supply Chain Management differs from traditional efforts to manage resource procurement, manufacturing and the delivery of products in at least two ways that represent dramatic shifts in the paradigms used to address these issues. First, the supply chain is seen as a single interdependent process not as isolated functions controlled by independent individuals, divisions or companies. According to Waller, "Supply chains operate across functions in an organization, across players in the chain and across national boundaries."[2] Second, all members of the supply chain (both internal and external) are seen as sharing a vested interest in the ultimate success of the chain in meeting the needs and expectations of customers.

The focus of SCM is to achieve the results needed to survive in an environment of increased competition. In addition to the traditional factors of cost and technical excellence these include: service, enhanced quality, reduced response time (cycle time), less inventory, and continuous innovation. Components of SCM include: customer service — to identify customer needs and expectations; product design, purchasing, manufacturing; distribution, product handling and delivery.

Why Is Supply Chain Management Needed?

Supply Chain Management is needed because the competitive environment of business organizations is changing and the targets for improvement are becoming more limited.

In their recent book *The 21st Century Organization,* Bennis and Mische describe the historical business environment in the U.S. by ten factors that were relatively orderly and predictable.[3] The table on the following page identifies the factors and provides you with the opportunity to assess their viability in the current economy and in your field of business. Those of you whose assessment indicates your belief that the historical competitive environment still exists may decide that supply chain management is not a relevant issue for you. However, you may do so at some risk because in the majority of companies, the role of SCM is increasing in importance.

Bennis and Mische clearly believe that change is taking place and they have developed predictions about the reality of the competitive arena in the 21st century. These predictions involve twelve trends:

1. Global markets will become saturated.
2. Technical advantages will be short-lived.
3. Service will be critical.
4. Growth in disposable incomes will slow.
5. Income gaps will widen.
6. Europe will achieve economic unification.
7. The Pacific Rim will hold the greatest potential for growth.
8. Technology will be the great equalizer.
9. The work force will be transitory.
10. Social enclaves will emerge in the U.S.
11. Economic boundaries will be transparent.
12. Breakthrough performance will only be achieved by making use of intellectual assets.[3]

Environmental Characteristics of the U.S. Economy, 1996

Rate each item in terms of its viability for the U.S. economy and your particular business sector. (1 = high application, 2 = moderate application, 3 = some application, 4 = little application, 5 = no application)

		U.S. Economy	*Specific Sector*
1.	High market growth	1 2 3 4 5	1 2 3 4 5
2.	Continued economic expansion	1 2 3 4 5	1 2 3 4 5
3.	Strong nationalism	1 2 3 4 5	1 2 3 4 5
4.	Massive migration of European cultures to the U.S.	1 2 3 4 5	1 2 3 4 5
5.	The standardization of products and manufacturing techniques	1 2 3 4 5	1 2 3 4 5
6.	Increasing personal income	1 2 3 4 5	1 2 3 4 5
7.	Simplification of work to its smallest elements	1 2 3 4 5	1 2 3 4 5
8.	Homogeneous markets	1 2 3 4 5	1 2 3 4 5
9.	Strict chains of managerial control	1 2 3 4 5	1 2 3 4 5
10.	The vertical integration of the enterprise	1 2 3 4 5	1 2 3 4 5

Adapted from Bennis and Mische, 1995, pp. 23–26.

All of these trends have important implications for the future of business organizations but several needed to be highlighted because we believe they are directly related to the emerging need for and value of supply chain management. These are:

1. **Global markets will become saturated.** The marketplace will be glutted with products, services, and suppliers. Any strategic advantage gained through a product will be temporary, because that product will be copied quickly and easily.
2. **Technical advantages will be short lived.** Information technology will be replicated quickly; thus, any significant competitive advantage obtained through technology will be quickly neutralized. Information technology will enable the work force to spread out globally, even though the enterprise is consolidated or headquartered at a specific location.
3. **Service will be critical.** As consumers become more knowledgeable, better educated, and more discriminating, markets will become more competitive. The service and experience that the consumer has in the relationship with the service provider will be the differentiating factors.
4. **Growth in disposable incomes will slow.** The markets for consumer products and durable goods will become tighter and more competitive as the growth rate in disposable income and savings continues to slow.
5. **Europe will achieve real economic unification as it continues to integrate.** Thus the benefits of large scale industry often attributed as a factor in U.S. economic growth will also stimulate European economic expansion.
6. **The Pacific Rim will hold the greatest potential for growth.** Continued double-digit growth in countries such as Korea, Taiwan, Singapore and revived economic activity in China will serve as a catalyst for global expansion for most U.S. industries.
7. **Technology will be the great equalizer.** Service and the ability to adapt to change rapidly will be the key success factor in the 21st century. To be competitive, an organization will have to be technology-enabled. Its people will use technology not just to perform transactions, but also to analyze information to support the decision-making process. Applications of data will be accessible, targeted to specific actions to quickly react to stretch goals in a boundaryless basis.

8. **Economic boundaries will be transparent.** As organizations strive to cultivate new and existing customers in increasingly competitive markets, provide service to those customers, optimize manufacturing and supplier resources, identify emerging opportunities, and prioritize investments geographical demarcations will be essentially nonexistent.
9. **Breakthrough performance will only be achieved by making use of intellectual assets.** There is no other way to generate the kind of performance that will carry an organization to success in the 21st century.[3]
10. **Overall, these trends will impact the competitive environment by:**
    - Increasing the globalization of the economy.
    - Increasing competition.
    - Increasing demands for value — speedy response time, service, quality.
    - Increasing reliance on information and analysis .

    In turn, these changes in the competitive environment will impact each of the dimensions of market attractiveness identified by Porter:

    - Ease of entry of new competitors
    - The threat of substitutes
    - Bargaining power of buyers
    - Bargaining power of suppliers
    - Intensity of rivalry among existing competitors[4]

Supply chain management provides a mechanism for addressing many of these issues for at least two reasons. First, in many organizations the opportunities for additional internal improvements are limited because of the implementation of quality improvement and reengineering efforts. Second, it provides an opportunity to apply many of the principles and techniques associated with these approaches to a new target — the external environment. Observations concerning the need for and the benefits of Supply Chain Management include:

U.S. corporations (at least the most successful) have done nearly all that can be done to optimize operations within the boundaries of their organizations. If they wish to continue to improve they must look outside of the traditional boundaries to the total supply chain. This means improving the effectiveness and efficiency of their logistics operation. The objective is simple; to boost competitiveness and profitability.[5]

Product excellence no longer provides a guarantee of competitive advantage. To develop and maintain such an advantage companies must use supply chain partnerships to reduce costs and complement their products with basic value-added relationships. Becton Dickinson & Co. derives a significant competitive advantage from supply chain management (SCM). The company evaluates its supply chain program against strategic imperatives involving productivity, quality, service development and supply chain integration.[6]

To remain competitive in the face of global competition and mounting cost pressures, firms must view the supply-chain as an integrated enterprise of closely-linked locations, with the flow of products along the supply chain driven by demand.

The opportunity to reduce costs and speed this product flow comes not only from improving the process at the locations, but also from integrating the information about demand and supply prior to the physical product flow.[7]

Implementing Supply Chain Management

As indicated above, Supply Chain Management differs from traditional efforts to manage resource procurement, manufacturing and the delivery of products in at least two ways: first, the supply chain is seen as a single interdependent process not isolated functions controlled by independent individuals, divisions or companies; and second, all members of the supply chain (both internal and external) are seen as sharing a vested interest in the ultimate success of the chain in meeting the needs and expectations of customers.

To produce the desired results Supply Chain Management ultimately requires the creation of an integrated supply chain system from design through raw materials to deliver to the end user — including support services and third party suppliers where appropriate.

The first condition needed for the development of a Supply Chain Management effort is the realization that the supply chain is a system. This means recognizing: (1) the interdependencies between members of the system both in terms of operations and goals, and (2) that an integrated supply chain will benefit all members of the system. After all, we talk about supply chains because we at least rhetorically recognize the reality and implications of linkages. Once these realities are recognized members of the system are in a position to apply the techniques of systems-process thinking and cross-functional collaboration. They are also in a position to make decisions for the benefit of the entire system not just their individual organizations. Leaders in the field of SCM, Becton Dickinson for example, speak in terms of partnerships. It must be emphasized however, that achieving a truly integrated system of this caliber requires high levels of mutual credibility and trust — far more than normally experienced in the relationships between supply chain members.

However, merely recognizing the need for integration does not produce an integrated supply chain — other conditions have to be achieved. These conditions include: (1) increased connectivity between members of the supply chain; (2) alignment of inter-organizational support systems; and (3) sharing of resources, information and expertise.

Connectivity

In order for SCM to work coordinated activities and planning between members of the chain are necessary. The goal of SCM is not merely to improve the operational flow of materials and services through each link in the chain. The ultimate goal is to develop an integrated system that allows members to anticipate the emerging needs of members of the system and to coordinate their efforts in the development of products, services and markets.

In most cases this means the nature and quality of the relationships (connections) between all members of the supply chain (suppliers, customers, third-party service providers, etc.) must be targeted for improvement. At a minimum improvement will require: (1) greater coordination of planning and implementation for business and operational activities; (2) the alignment of strategic logistical activities that impact the supply chain; (3) the alignment of structural factors that impact the supply chain — particularly the organization of human and physical resources; (4) a mutual understanding of the goals and expectations of member organizations — both for their individual organization and for the entire supply chain.

The importance of the inter-organizational relationships between members of a supply chain has recently been recognized in one of the concluding propositions in a study of global logistics best practice. Labeled as important, this principle states that there should be an emphasis on establishing and maintaining strong value-added relationships.

- Leading-edge firms have clear policies and procedures related to alliance establishment, management, modification and termination.
- Successful alliances range from casual to highly complex relationships. All permeate organizational boundaries in a meaningful way.

- While information technology is important to successful relationships, it is not prerequisite if there is a willingness to share information.[8]

Inter-organizational Support Systems

In order to enhance the productivity of the entire chain, Supply Chain Management requires increased alignment of the support systems that impact inter-organizational boundaries. As shown in Profile 4.1 and Exhibit 2, specific support systems that may require higher levels of integration include:

1. **Information Systems** — hardware, software, and communication systems
2 **Performance Measurement Systems** — accounting, financial, logistical (purchasing, inventory, delivery, etc,
3. **Human Resource Systems** — incentives, rewards and recognition, productivity, etc.

Compatible and preferably integrated information systems are needed for units of the supply chain to communicate effectively. This communication process helps (1) reduce translation steps; (2) reduce the potential for miscommunication; (3) enhance the potential for the electronic dissemination and receipt of business information. Compatible information systems will also enhance the ability of supply chain members to share in the analysis of aggregated data and address systemic issues rather than focus on specific orders and/or problems as shown in Profiles 4.1 and 4.2.

The alignment of measurement systems is needed to create common metrics for tracking and assessing organizational activities. The alignment of measurement systems has value for at least two reasons: first, and similar to the need for comparable information systems, the alignment of measurement systems helps ensure that members of supply chain have a common metric and performance measures — short a common language. This is an important methodology for tracking logistical activities and in assessing the performance of the supply chain on varying criteria such as financial performance. Second, the use of a common metric helps emphasize common goals held by members of the systems and the overriding goals of the supply chain itself. Performance measurement is also recognized as an indicator of global best practice and is increasing in scope and importance.

- Measurement focus is moving from traditional focus on cost, customer service, asset utilization, productivity and quality to encompass and emphasize overall supply chain performance.
- Emphasis is on comprehensive measures that calibrate relevancy and responsiveness.
- Control and information concerns are focusing on zero defect performance required for perfect order performance.[6]

The alignment of human-resource policies and procedures (incentives, rewards and recognition) is needed to help insure that members of the chain are motivated by similar goals and evaluated by similar assessment systems. The achievement of this condition is important because humans are and will continue to be a critical resource in all organizational activities including Supply Chain Management. Also, reward and recognition systems are slowly being revamped to incentivize meaningful work.

- Easily the most complex issue facing implementation of advanced logistical concepts.
- General sense of managerial uneasiness about rate of advancement and increased perceptions of declining loyalty.
- Requires meaningful dialogue throughout total organization creating a new general management challenge for logistics executives.[8]

Sharing of Resources — Information and Expertise

As indicated above, the competitive environment of the 21st century will be characterized by (1) increased globalization of the economy; (2) increased competition; and (3) increased demands for value including speedy response time, service, and quality. To meet these conditions and expectations, the performance of the total supply chain must continually improve. Sharing resources between members of the supply chain will be a major vehicle for insuring continuous improvement at both levels — particularly the sharing of information and expertise. In short, achieving a state of continuous improvement means all supply chain members will have to respond to problems in any link of the chain (or between links in the chain) as if it were their own problem — because ultimately they will be.

Successful implementation of Macrologistics-based supply chain management will require dramatic shifts in the paradigms, structures, policies and behaviors currently used to address to address supply chain issues. Shifts need to occur because:

- Company structures are based on a philosophical foundation inherited from Adam Smith that emphasizes the belief that maximizing the performance of a system is best achieved by encouraging members of the system to maximize their individual interests and their independence.
- Existing organizational units emphasize the "silo" mentality characteristic of the functional structure employed in the internal management of most organizations. Among other negative effects, this leads to the minimumization of information and resource sharing within organizations let alone across organizational boundaries. Silos need to be broken down so that cross-functional within organizations and cross boundary cooperation exists with members of the external environment — expand contact with customers and suppliers.
- There is a need to emphasize a pooled approach to achieving the common good (adding together the performance of individual units) as opposed to an integrated and synergistic approach that emphasizes interdependence and reciprocity.
- A paradigm shift is necessary to encourage the independent development and implementation of business strategies, products and/or services and operational activities. (See Profile 4.2.)
- It will be helpful to identify the processes used to design information systems that often results in incompatible hardware, software and communications systems. These forms of incompatibility and inconsistency need to be connected to permit effective design of performance measures.
- Existing definitions of organizational units do not easily provide the mechanism for or the tradition of developing the credibility and trust needed for meaningful dialogue and negotiations.
- The structures are based on a legal tradition that emphasizes the separation rather than the integration of supply chain members. These legal structures hinder the implementation of SCM efforts and need to be replaced.

Activities to Initiate and to Facilitate Supply Chain Management*

There are no simple cookbook methods to develop an effective Supply Chain Management effort. As indicated above, the first step is the realization that the supply chain is a system and that all stakeholders in the system can benefit if it is managed to achieve greater integration. The direct causes of this realization are ultimately twofold: (1) some form of discomfort or threat to existing arrangements; and (2) a vision of a better future.

Whatever the reasons, supply chain members must come to the belief that change toward a more integrated system is both desirable and necessary. Once this condition is achieved supply chain members will intuitively move towards more integrative behavior. However, a number of specific activities can be initiated to help create an environment that fosters the implementation of Supply Chain Management efforts. Figure 4.1 and Profile 4.1 provide diagrammatic summaries of these activities and their relationship to each other.

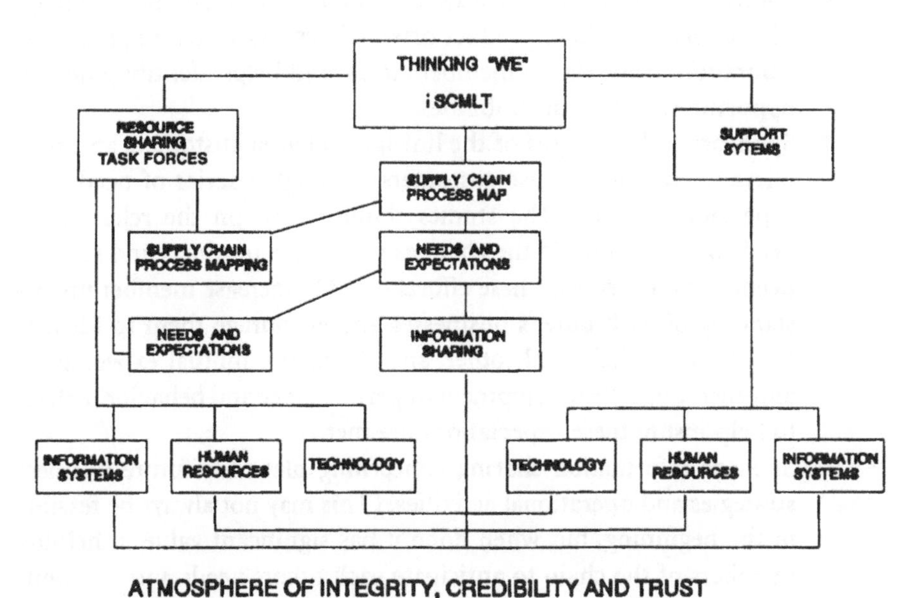

Figure 4.1 Implementing Supply Chain Management

*Courtesy of Ralph Lewis.

Connectivity: Thinking "We"

Efforts to increase the perception and reality of connectivity must be built on and support the development of an atmosphere of integrity, credibility and trust. Such an atmosphere does not emerge full grown like Athena from the head of Zeus. Rather, it evolves from ongoing interactions focused on relevant issues. These interactions provide the opportunity for information focus, dialogue and team building needed to develop the behavior and values required for effective Supply Chain Management. Specific activities may include:

- The creation of an Inter-Organizational Supply Chain Management Leadership Team (IOSCMLT).The role of this team is to focus attention on the supply chain (the we) and the overall needs of the chain — not solving specific cases/issues. In turn the ISCMLT may sponsor the creation of problem oriented inter-organizational problem solving teams.
- Sponsoring the creation of a systems level supply chain process map. This effort provides a product oriented activity which requires collaboration, helps clarify members roles, and helps identify potential opportunities and vulnerabilities.
- Sponsoring the analysis of the linkages between customers and suppliers (including end use customers) through a series of needs and expectation studies. The studies should focus on the relationship needs between links in the chain as well as the products and services needed and expected. These efforts will: (1) increase member understanding of each other's businesses and encourage them to identify how they can help each other, and (2) clarify mutual expectations and encourage the development of performance and behavior metrics to help insure these expectations are met.
- Sponsor information sharing concerning plans for future business strategies and operational activities. (This may not always be feasible in the beginning, but when done it has significant value in helping members of the chain to anticipate each others needs and to simultaneously develop products and services. Specific activities may include reciprocal site visits and the inter-organizational dissemination of mutually relevant materials.

Inter-organizational Support Systems

Efforts to develop compatible and integrated support systems may include:

- The development of a Supply Chain Information System (SCIS). The first steps in this process should be an information inventory that includes all members of the chain and focuses on: (1) information systems; (2) performance measurement systems; and (3) human resource systems. Inventory questions should include as:
  What data is available?
  In what form do we have it?
  Who has it?
  How can it be accessed?
  What data do we need?
  How will we use it?
  How can we get what we need but don't have?
  What information systems (hardware, software, communications) do we use?
  What plans do we have for enhancing our information systems.
- The development of a Supply Chain Human Resource Inventory. This inventory should identify: (1) individuals and teams with specific expertise relevant to the supply chain, and (2) chain members with special human resource needs.
- The development of a Supply Chain Technology Inventory. This inventory should identify special equipment, technology, systems available within the supply chain and technology that can be acquired from outside resources (e.g., bar coding, software, logistics software providers and other organizations.

Sharing of Resources — Information and Expertise

Initial efforts to share resources could include:

- The establishment of an Inter-organizational Task Force to develop a systems level map of supply chain processes.
- The establishment of an Inter-organizational Task Force to conduct the customer-supplier needs and expectations studies.

- The development of additional Inter-organizational Task Forces for the: (1) Supply Chain Information Systems Inventory; (2) Supply Chain Human Resource-Expertise Inventory; and (3) Supply Chain Technology Inventory.

The cases below provide an illustration of Supply Chain Management as employed by Becton Dickinson, Nippon Steel, and MicroAge (see Profiles 4.1, 4.2, and 4.3). These efforts are customized to the needs of each company and represent "prototype" efforts at the cutting edge of this new management tool. Since the methodologies are in their formative stages, substantial room exists for innovative expansion of these concepts.[10]

References and Endnotes

1. Painter, Jim, Integrated Logistics — The Way Ahead Evolves, Purchasing & Supply Management, Logistics Supplement June, 1994.
2. Waller, Alan, Supply Chain Management, *Across the Board*, Vol. 32, No. 3, March 1995.
3. Bennis, Warren and Mische, Michael, *The 21st Century Organization: Reinventing Through Reengineering*, Pfeiffer & Company, San Diego, 1995,
4. Porter, Michael, E., *Competitive Strategy: Techniques for Analyzing Industries and Competitors*, Free Press, New York, 1980.
5. Copacino, William C., A New View of Logistics, *Traffic Management*, Vol. 32, No. 12, Dec. 1993.
6. Battaglia, Alfred J., Beyond Logistics: Supply Chain Management, Chief Executive, No. 99, Nov/Dec 1994.
7. Vonchek, Arthur, The Components of Supply Chain Management, *Logistics Focus*, Vol. 3 No. 3, April 1995.
8. Bowersox, Donald J. et al., Global Logistics Best Practice: An Intermediate Research Perspective, Annual Conference Proceedings, Council of Logistics Management, Cincinnati, Ohio, October 16–19, 1994.
9. Copacino, William C., *Supply Chain Management: The Basics and Beyond*, St. Lucie Press, Boca Raton, FL, 1997.
10. For the application of these principles to small-business organizations, see the work of Michael Fedotowsky, President of Customer Focused Technologies, a Dayton-based consulting firm. Mr. Fedotowsky is a pioneer in the application of quality and logistics concepts to small-business organizations.

Becton Dickinson: Value-Added Supply Logistics to Achieve Competitive Advantage*

Mission/Vision at Becton Dickinson, Superior Quality Is the Only Way

Our mission as a company is to provide the many markets we serve with products of consistently superior quality at price levels that are fair and competitive. Achieving this mission is a responsibility that we all share and is necessary to meet the expectations of our customers, ourselves, and our shareholders. With this uncompromising dedication to superior quality, we have a focus for our actions that unifies us, adds value to our work, and enriches our lives.
Address: Becton Drive, Franklin Lakes, NJ 07417-1880
Industry Category: Medical Products and Services
Corporate Description: Becton Dickinson and Company manufactures and sells a broad range of medical supplies and devices and diagnostic systems for use by health care professionals, medical research institutions, and the general public at large.
Size Revenues: Approximately $2.8 billion annually.
Number of Employees: About 19,000 as of 1994.

Background Information

According to Nicholas LaHowchic, VP of Logistics Operations, the Becton Dickinson logistics story began with the understanding that having the best product was not enough to sustain competitive advantage. Their customers also expect services such as perfect delivery, accurate invoicing, perfect installation, JIT training, technical consultation, responsive field reps, timely information, and much more. Realizing that fundamental changes were required, and that the 1990s were not a time to implement fads, they began their search for the ideal change management solution — the Supply Chain Management (SCM) component of Macrologistics Management.

During the 1980s, the leadership team realized that "getting out of the box was the role of leadership." They studied the lessons of 43 companies identified in the book by Peters and Waterman *In Search of Excellence* and found that between 1982 and 1988, 66% had lost competitive position. The lesson was

* © Strategy Associates, Inc. We are indebted to Nick Lahowchic for Becton Dickinson source materials.

obvious: it was easier to achieve "world class" status than to maintain it. They realized that fundamental changes were required to gain and maintain competitive advantage, and that change management needed to be on the company's agenda.

Introduction of Supply Chain Management

The Becton Dickinson strategy for sustained profitability became Supply Chain Management which they define as follows:

> "An integrating process, used to build competitive position, based on the delivery to customers of basic and unexpected services. Led by line executives, SCM optimizes information and product flows from the purchase of raw materials to the delivery of finished goods with a vision of achieving significant strategic objectives involving productivity, quality, innovative services, and alliances. Total Supply Chain Management includes the implementation of sales and marketing activities to share the benefits with all of the participants in the supply chain."

The Macrologistics management implementation model used at Becton Dickinson involved five interrelated strategies centering on the management of the supply chain:

1. Moving from a department focus to a process integration
2. Eliminating process steps instead of merely speeding up the process
3. Developing customer-driven thinking instead of "inside-out" thinking
4. Building productive relationships instead of conducting adversarial negotiations
5. Moving from price to total cost, as shown in Exhibit 1.

Conditions for Effective Change

The Becton Dickinson Macrologistics management team identified seven conditions for effective change:

- Condition #1: We must want or need to make a change.
- Condition #2: We must have a vision of where we would like to be.
- Condition #3: We must be reasonably satisfied that the benefits will be greater than the costs and difficulties of making the change.
- Condition #4: We must have an appropriate implementation model which provides a means of moving from where we are to where we would like to be.

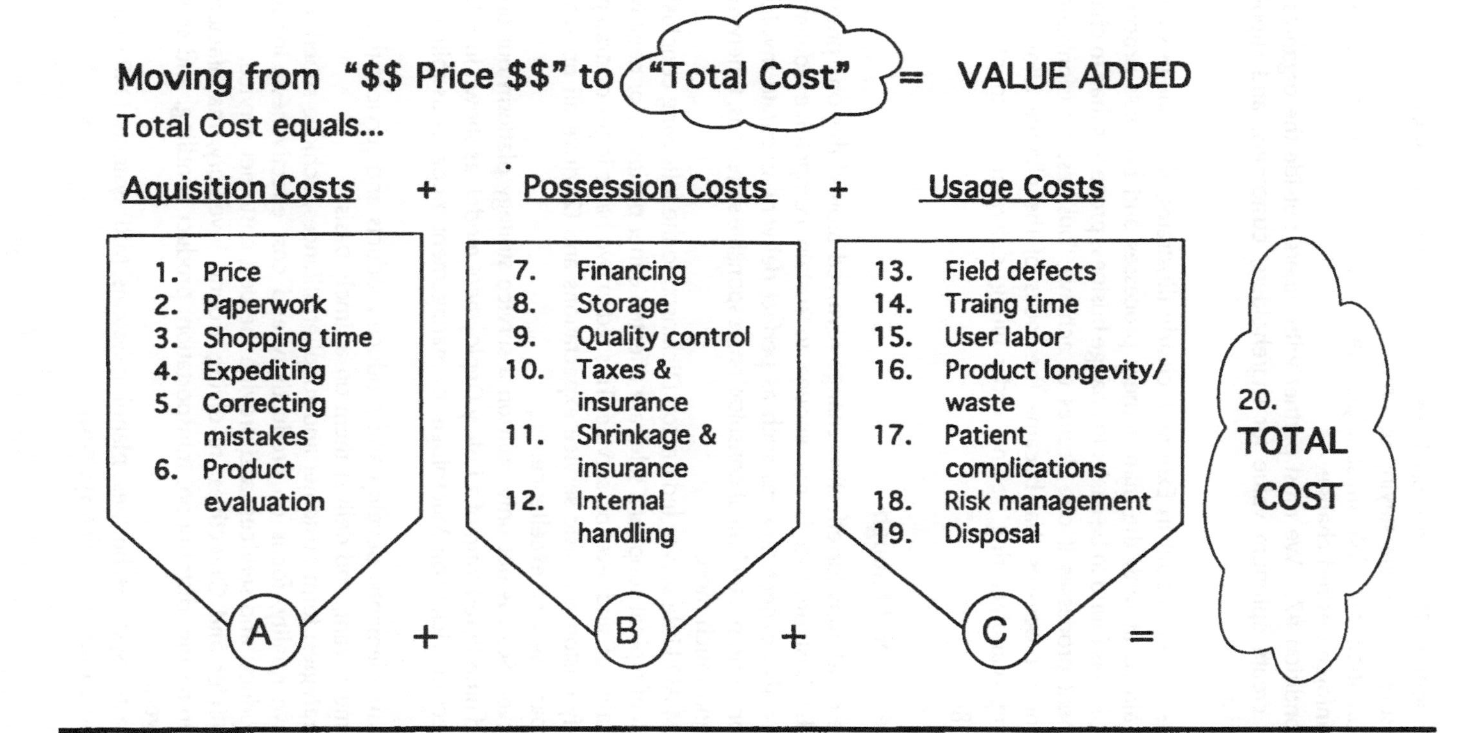
Moving from "$$ Price $$" to "Total Cost" = VALUE ADDED
Total Cost equals...
Aquisition Costs
+
Possession Costs
+
Usage Costs
1. Price
2. Paperwork
3. Shopping time
4. Expediting
5. Correcting mistakes
6. Product evaluation
7. Financing
8. Storage
9. Quality control
10. Taxes & insurance
11. Shrinkage & insurance
12. Internal handling
13. Field defects
14. Traing time
15. User labor
16. Product longevity/ waste
17. Patient complications
18. Risk management
19. Disposal
A
+
B
+
C
=
20. TOTAL COST

Exhibit 1 The Becton Dickinson Value-Added Model

- Condition #5: We must have a measurement monitoring system to assure that we are staying on course.
- Condition #6: We must adjust reward and recognition systems to reinforce desired change.
- Condition #7: We must partner with others outside the organization to create optimum value for shareholders, customers, and suppliers alike.

As the change model in Exhibit 1 clearly illustrates, the change strategy must be aimed at both the management processes and business processes. However, it was found to be easier to change business processes than to change management processes if one desires to achieve "business transformation". The business processes that became the focus of the change effort were: developing, manufacturing, selling, order filling, delivering, accounting and organizing.

When Best Is Not Enough

All of the conditions for effective change revolved around this one premise: Having the best product is not enough to sustain competitive advantage. Customers also expect services such as perfect delivery, accurate invoicing, installation, training, technical consultation, responsive sales reps, timely information, and much more.

A 1994 AD Little survey led Becton Dickinson to the following conclusion: 5% to 6% of typical supplier's sales are affected either positively or negatively by the quality of, and level of, service. This led to two basic logistics strategies: (1) identify customers basic service expectations and (2) drive all operations towards basic service excellence.

However, SCM is not only built on a service strategy platform but on an integrated nine-tiered form of a Policy Deployment model, as shown in Exhibit 2. The Becton Dickinson Macrologistics management model is as follows:

- **Basic Elements:** develop and produce products and services that customers want, and deliver them on a timely basis.
- **Strategies:** form customer and supplier alliances, achieve superior relative quality, focus on productivity and cost effectiveness, innovate products and services, and provide the best customer service.
- **Policies and Objectives:** revolving around inventory, manufacturing, purchasing, distribution, transportation, product handling, and service levels.
- **Master Systems:** business planning systems, MRP systems, DRP systems, and customer service systems.

- **Sub-Systems:** including forecasting, capacity planning, production and vendor scheduling, inventory status, MRP, warehousing management, lead-time management, logistics maps, and order management.
- **Staff/Skills/Style:** assure adequate staff resources, build required skills, and create a multi-functional integrated teamwork approach.
- **Measurements:** supplier performance, manufacturing performance, capacity utilization, inventory investment, cost effectiveness, service levels, and customer satisfaction.
- **Shared Values:** implement World Class supply chain management in the health care industry.
- **Benchmarking:** against the best-in-class practices.

Another unique feature of the Becton Dickinson Macrologistics Management model is their approach to partnering with their distributors through a "partnership platform" as follows:

- **Information Technology Services:** including several basic types of SPEED-COM platforms (a) order entry (b) advanced shipment notification/electronic invoicing (c) electronic contract notification (d) electronic rebate (e) net billing.
- **Logistics Services:** including (a) rapid service distribution centers (b) consolidated shipments (c) scheduled delivery (d) guaranteed delivery.
- **Shared Platforms with Distributors:** including SPEED-COM LINK II automatic replenishment system, and shared information systems, such as EDI and others.

Nippon Steel Corporation: How to Manage Innovation — 4.2

The steel industry faces many challenges. It is a mature industry with more capacity than demand, and its products are being replaced by plastics and other synthetics. As a result of competition from other heavy industries and a downturn in the economy, many steel producers have gone out of business. Despite this unfavorable environment, Nippon Steel of Japan has persevered. Its competitiveness stems from its ability to innovate.

The following discussion explores Nippon's success from four perspectives: the challenges the company faces, its approach to technological innovation, its emphasis in innovation management, and the lessons that can be learned from its experience.

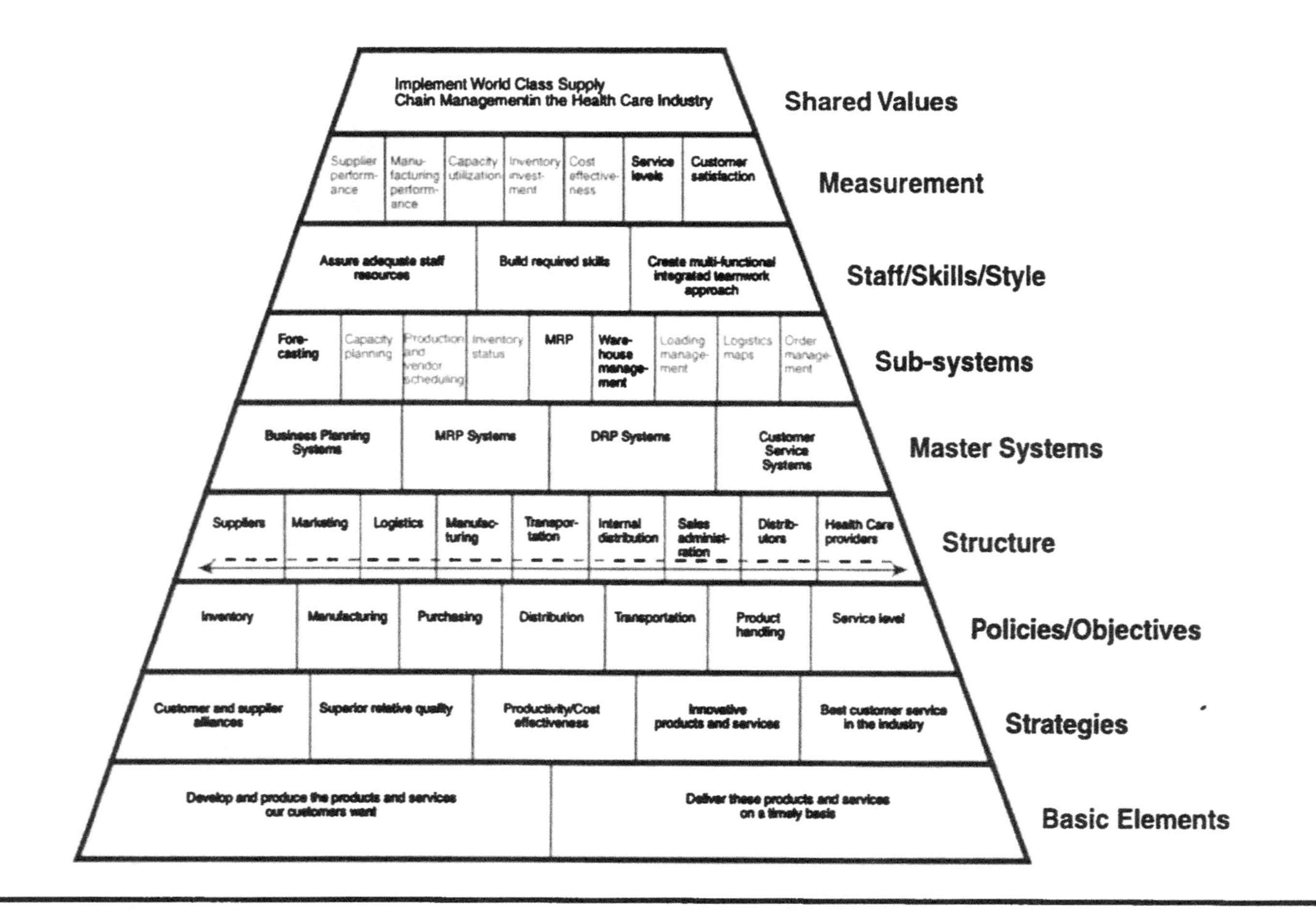

Exhibit 2 The Becton Dickinson Macrologistics Management Model

According to Nippon Steel's management, several major challenges result from the current business environment. They are:

1. The need to introduce new products and improve manufacturing processes.
2. The need to integrate economic functions, such as marketing, design, and operations, so as to speed product and process innovation.
3. The need to scan the environment for future opportunities and threats, and to use new technologies before others do.
4. The need to introduce entirely new technologies.
5. The need to develop new management systems to cope with rapid change.
6. The need to change organizational structure and information management systems, and to speed up the CPS process because of the shortening cycle time between design and manufacturing.
7. The need for specialization as well as integration.
8. The need to recover R&D costs.
9. The need for continuous skill development to match new innovations.

In its efforts to deal with these challenges, Nippon Steel considers innovation the most critical aspect of its strategy.

At the center of Nippon Steel's innovation strategy is technological innovation. The company has grouped its technological innovations into seven categories:

1. Integrating different steps of a process in order to reduce processing delays or waste, or drastically reduce processing time. *Example:* linking continuous casting and direct rolling into an integrated process.
2. Developing new products based on market demand. *Example:* fire-resistant structural steel.
3. Developing new processes and instruments to increase quality, improve a process, or make monitoring systems more effective. *Example:* artificial intelligence controls for blast furnaces.
4. Assimilating acquired technology and transferring it horizontally into new fields. *Example:* gold bonding wires.
5. Developing new construction or fabrication methods. *Example:* air-inflated double-membrane stainless-steel roof.
6. Creating new technologies by fusing two diverse technologies. *Example:* laser neural network to detect defects in rolling.
7. Using joint ventures to develop new technologies and integrate competencies in diverse fields. *Example:* satellite broadcasting receiver system.

Innovation management at Nippon Steel is typical of that found in many Japanese firms. It consists of the following elements:

1. **Forward looking** — to identify new business areas.
2. **Organizational intelligence** — about competitors, new technologies, and other environmental factors.
3. **Four types of organizational learning** — maintenance, adaptive, transitional, and creative.
4. **Technology fusion** — merging of two technologies to form a new technology.
5. **Concurrent engineering.**
6. **Competence building** — through joint ventures.
7. **Horizontal information flow** — using a structure specifically created to promote functional integration.
8. **Intensive skill development in all areas** — reinforced by organizational learning and horizontal information flow.

Nippon Steel's innovation strategy is set forth in the following statement by B. Bowonder and T. Miyake:

> "Organizational changes are needed for responding to market changes in its place. New information and communication technology systems have to be implemented, which can facilitate organizational communication for coping with the dynamic changes in the environment. Implementing new information and communication systems can simultaneously support a greater degree of centralization and coordination, and at the same time promote flexibility. New information and communication technologies will stimulate innovations through a variety of modes such as:
>
> - Integrated organizational intelligence made possible by rapid communication and data exchange.
> - Rapid decision making facilitated by information networking and functional integration.
> - Computer graphics, computer aided design and engineering, facilitating design and engineering sharpening the skills.
> - Technical decision support systems, expert systems, and supercomputers, permitting knowledge-based real-time controls and competence fusion.
> - Real-time information exchange through high-quality facsimile system through digital transmission.
> - On-line technical information support system for getting ideas.

Nippon Steel Corporation, as a creative integrator and as an innovative supplier, is expanding its business into new technologies, coupling its own assets with external strengths through original development, establishment of new firms, and capital participation in or business ties with other firms. This is the essence of Macrologistics strategy at Nippon Steel Corporation.

The strategy adopted by Nippon Steel Corporation indicates that 'time' is the most crucial determinant of innovativeness. Through joint ventures, concurrent engineering, strategic alliances. networking, multifunctional new development teams, and new-product subsidiaries Nippon Steel has been able to bring out innovative product/process changes rapidly. Functional integration, together with a decentralized operation, is the organizational characteristic supporting innovations. The win-win cooperative strategy adopted by Nippon Steel, in the form of a variety of new subsidiaries with other firms as partners for competence fusion, has helped it in rapidly reaching the market. The forward-looking approach is supported by a variety of techniques for starting new business segments. The approach of Nippon Steel is highly interactive, with strong synergy in which the emphasis is on diffusion of innovations for developing new business."

 Source: B. Bowonder and T. Miyake, "Management of Corporate Innovation: A Case Study From the Nippon Steel Corporation," *Creativity and Innovation Management* (June 1992), pp. 75–86. Used with the permission of Blackwell Publishers, Oxford, UK.

Microage Computer Centers Innovates with Supply Chain Engineering 4.3

Microage Computer Centers is a $3 billion distributor of personal computers, software, and related products to franchise stores and customers throughout the U.S. The company relies on four basic logistics principles that help put its supply chain into place and make it more efficient: process mapping, building internal relationships, selling logistics to executive management, and the execution of breakthroughs and new ideas.

Mapping business processes involves identifying and flowcharting the organization's core and support processes. The order-cycle process was the target, as well as the physical distribution process throughout the channel. A team was formed with representatives from purchasing, receiving and shipping, order picking, inventory control, and risk management. Briefings were held with executive management to keep them informed as to future directions in the logistics field. This information was also used for strategic planning involving the logistics department.

After the processes are initially flowcharted, the carriers and logistics partners are included in the team meetings to discuss the value-added service that they are able to bring to help Microage become more competitive by integrating these services into the core competencies. The chief logistics officer has an integrated communications plan that includes everyone in the supply chain, in order to become an information rich organization and innovate for tomorrow.

Source: Council of Logistics Management, Annual Conference Proceedings, San Diego, California, October, 1995, pp. 335–349.

Value Chain Quality: A Ten-Point Model 4.4

In reading through the various articles written in the last several years about logistics quality management, most seem to be driven by a focus on one of three components of the value-chain quality process: (1) the importance of establishing and using standards; (2) the need for cross-functional logistics teams; and (3) the value of empowering employees. The author points out, however, that most approaches offer a singular method which, standing alone, do not provide an appropriate long-term solution for all organizations. What is lacking is a flexible model that allows each organization to tailor the process to its unique circumstance and serve as a guideline for future action.

Creating and maintaining organization-wide effectiveness requires an integrated logistics approach involving the ten key components listed below. While improving any single component may lead to better performance, a comprehensive organizational transformation requries action on all ten:

1. Establish a culture of logistics quality
2. Develop a team orientation
3. Develop leadership skills
4. Develop customer-driven policies and procedures
5. Set logistics standards
6. Develop human resources
7. Plan for logistics quality
8. Build systems to measure achievement
9. Evaluate performance
10. Build reward and recognition systems

The first three elements make up the informal components of an organizational response to quality management. They address the basic aspects of how people work together to accomplish their goals and objectives. The remaining seven deal with the organization's systems and procedures used in day-to-day logistics operations.

Source: Kenneth Heymann, *Cornell H.R.A. Quarterly,* October 1992, p. 51–60.

5 Information Systems and Technology

Information Technology and cost-service tradeoffs are two of the most powerful and definitive forces in the development of modern civilization and a key component of Macrologistics Management. When logistics managers plan to approach the cost-service issue, however, their typical approach is through the application of Information Technology. Thus, the correct identification, **alignment**, and implementation of information Technology is the single most important issue facing the industry today.

ENIAC

Commonly thought of as the first modern computer, ENIAC was built in 1944. It took up more space than an 18-wheeler's tractor trailer and weighed more than 17 Chevrolet Camaros. It consumed 140,000 watts of electricity while executing up to 5,000 basic arithmetic operations per second. One of today's popular microprocessors, the 486, is built on a tiny piece of silicon about the size of a dime. It weighs less than a packet of Sweet 'N Low and uses less than 2 watts of electricity while executing up to 54,000,000 instructions per second.

Business Week, "The Information Revolution"

Information Technology is no longer just computer hardware and software. Instead, telecommunications has become the driving force for most IT-based logistics opportunity. With telephone, PCs, Internet and television access to information and processing services, IT is becoming truly omnimedia. In other words, any information that is coded in digital form can be processed, stored and transmitted by any of these media. Overall, the following four building blocks are considered critical to fuel the Macrologistics Systems of tomorrow:

1. Information and service access tools.
2. Telecommunications links, including local area and wide area networks.
3. Processing engines, including linked complexes of machines and mainframes.
4. Information stores, including electronic libraries and multimedia storage devices.[1]

The Impact of Technology

Fueled by scientific development and discovery, the evolution of technology is synonymous with the history of man's struggle to improve conditions on the planet, the sheer will and desire to reach beyond the limitations of the day and create altogether new, unprecedented ways of travelling and transporting, communicating, working, creating and entertaining. Technology has transformed, and continues to transform, the way we live and move and think, and basically what we are.

Information Technology is not static. It moves in spurts and sometimes waves of discovery and innovation. Like a tornado, its course is often unpredictable and dangerous. Like shattering glass, it can break off in a thousand directions, spawning a huge networks of sub-innovations and spinoffs, each with their own unique shape, form and application. However, research studies continue to report that the pace of adoption of Information Technology in logistics is lagging behind the arrival rate of commercially available technology and the overall business adoption rate, as reported in a recent survey conducted by Golden State University.[2]

Discoveries, advances and innovations often merge and align to engender new applications, new products that would not be possible without the important contribution of each separate technological development. Guns, for instance, would never have come into existence if gunpowder, formerly used by the Chinese for fireworks, was not combined with the matchlock, the flintlock and subsequent advances in gunpowder-ignition mechanisms. Electricity, without developments in generation, transmission and distribution, would never have found its way into almost every household in the developed world. And oil, the precious fuel of the industrial age, would never have become so vitally important apart from the combustion engine, would have never even been extracted in mass quantities apart from drilling technologies, would never have been refined without processing technologies and would never have been consumed on a mass scale apart from the literally thousands of parallel technological developments undergirding the commercialization of energy and oil-powered transmission and travel.[3]

So it is true for the world of Macrologistics Management. In architecture and construction, travel and transportation, tools and devices, agriculture, medicine and communications, industrial-age advances are being made at a pace and on a scale never-before witnessed by human life. What we see with the advent of industry, in short, is an acceleration of change that is explosive beyond comprehension. However, of the 31 information technologies surveyed by Golden State University, only the following handful were being utilized by more than half of the survey participants: bar coding, fax, and all three computer processing platforms (mainframe, mini, and micro).[4]

Emerging Trends

Technologies with strong future possibilities much greater than current usage are: hand-held data entry, WORM/CD-ROM, bar coding, pen note pads, voice recognition, and EDI. The overall findings of the study were that logistics was utilizing only 18% of commercially available technologies in 1995. In fact, few organizations are experimenting with radio frequency, voice recognition, multi-media, and expert systems.

Yet in all their wondrous glory, the techno-developments that have been discovered before the 1950s are what futurist Alvin Toffler calls "second-wave" innovations — foundational and key elements of industrial society. They do not approximate or come close to the technological explosion of the information era, the third great wave of human history that has been forming slowly since the 1950s and at blinding speed since for the past ten years or so.

It is on these information-age developments that Macrologistics Management takes focus — how they shape and define the logistics environment, how they literally structure and restructure whole industries and businesses, how they change the structure of logistics competition, how they effect the way we manage, the decisions we make and the operating environment in which we make those decisions. From the way we plan to the way we communicate to the way we hire, train, design, purchase, produce, deliver and distribute — technology and logistics work hand-in-hand to achieve organizational objectives and priorities. Without the Information Technology, Macrologistics breakthrough may not be possible. In many cases, it is the same technology organized differently, as in the case of the Federal Express Memphis hub.

Already, the world is well into what futurist Alvin Toffler calls the third great phase of human history — the information age — about which he eloquently penned the following words in *The Third Wave*:

> "For Third Wave civilization, the most basic raw material of all — and one that can never be exhausted — is information, including imagination. Through imagination and information, substitutes will be found for many of today's exhaustible resources — although this substitution, once more, will all too frequently be accompanied by drastic economic swings and lurches.
>
> With information becoming more important than ever before, the new civilization will restructure education, redefine scientific research and, above all, reorganize the media of communication. Today's mass media, both print and electronic, are wholly inadequate to cope with the communications load and to provide the requisite cultural variety for survival. Instead of being culturally dominated by a few mass media, Third-Wave civilization will rest on inter-active, de-massified media, feeding extremely diverse and often highly personalized imagery into and out of the mind-stream of the society.
>
> Looking far ahead, television will give way to "indi-video" — narrowcasting carried to the ultimate: images addressed to a single individual at a time. We may also eventually use drugs, direct brain-to-brain communication, and other forms of electrochemical communication only vaguely hinted at until now. All of which will raise startling, though not insoluble, political and moral problems.
>
> The giant centralized computer with its whirring tapes and complex cooling systems have already been supplanted by myriad chips of intelligence, embedded in one form or another and will eventually be found in every home, hospital, and hotel, every vehicle and appliance, virtually every building block. The electronic environment will literally converse with us."

The Productivity of Knowledge Links

Information and its byproduct, knowledge, have firmly established themselves as the undisputed engines of the new economy. The spectacular productivity improvements of the past 50 years, in conjunction with the rapid rise of service- and information-intensive businesses and business functions, have caused tectonic shifts in the structure of the world economy that have hurled information into center stage and relegated labor- and capital-intensive businesses and business functions to a secondary, even tertiary, position in the economic pecking order.

Already, the presence and availability of information technology is transforming formerly labor- and capital-intensive industries such as farming and manufacturing. Italian businessman Vittorio Merloni, whose company makes 10% of all washing machines, refrigerators and other household appliances sold in Europe, says:

> "We need less capital now to do the same thing. The reason is that knowledge-based technologies are reducing the capital needed to produce, say, dishwashers, stoves or vacuum cleaners."

Among other Macrologistics applications, they are using information technology to reduce high-cost inventory and speed factory responsiveness to the market. The key is to use the same approach to bring these concepts to the entire supply chain to reduce the overall cost to the customer. Both Merloni and Fred Smith of FedEx saw opportunities for competitive advantage, using logistics as the key to expand market share.

In hundreds of thousands of companies, logistics-based information is assuming a greatly expanded role. Even in Germany, where manufacturing is still king, information technology has made its way into the center of business life — although apprentices at premier tool-maker Trumpf, still devote most of a year to hand-filing metal, learning to file metal by hand, one-third of the company's research staff is comprised of software engineers.

Even in the logistics sensitive grocery business, information technology has become a core and essential element, a prerequisite for survival. Through optical scanning technology, long checkout lines and errors in accounting have been minimized. The "Universal Product Code" or "bar code" — that small black box of lines and numbers that appears on everything from apple sauce to laundry detergent has forever changed the way products are packaged, distributed, stocked, sold and ordered. According to Toffler, bar coding has become nearly universal in the United States, with fully 95% of all food items marked by the distinctive little symbol. In France, 3,470 supermarkets and specialty department stores were using it. In West Germany, 1,500 food stores and 200 department stores employed scanners. From Brazil to Czechoslovakia and Papua-New Guinea, there were 78,000 scanners at work as far back as 1988.[5] Barcoding also enables Macrologistics Systems to align information by creating a database of specific product information available in a realtime mode.

But as with virtually any application of information technology, the examples cited represent only the infant stages of the information age, the first utterred expressions of a just-born technology. In the supermarket and retail business, for example, consumers may soon find themselves navigating their ways through isles lined with "electronic shelves." Instead of paper tags indicating the price of items, they will find blinking liquid — crystal displays with digital-price readouts. The implications are staggering. In addition to automatically changing thousands of prices from a remote location, the new displays would provide nutritional and other information at the touch of a button and even elicit market research information from the consumer.

In the world of Macrologistics retail, Wal-Mart is setting the electronic pace by requiring all its vendors to be tied into its system of electronic data exchange. Simply, through the use of integrated, computer-to-computer systems, Wal-Mart's suppliers know when the retailer is running low on specific products and send new inventory automatically — without any order being placed, without any unnecessary hand-offs or steps in-between. What we are moving toward is a seamless connection between consumers and manufacturers. According to George Fields, chairman and CEO of ASI Market Research (Japan):

> "Distribution no longer means putting something on the shelf. It is now essentially an information system. Distribution will no longer be a chain of inventory points, passing goods along the line, but an information link between the manufacturer and the consumer."

In this environment, "prosumers" — a term coined by Alvin Toffler in *The Third Wave* — become actively engaged in product design. Using CAD/CAM software, for instance, it may not be too far off in the future when prosumers will be able to participate in the design of their vehicles at the dealer's workstation. A prosumer can pre-select the body structure, drive-train components and suspension components. She can tailor the car's lighting system or instrument panel layout to fit her personal preferences. And, perhaps most impressive, her car can be delivered within three days after order is placed.[6]

In countless instances, across the board, information technology is redefining how companies conduct business and succeed in the marketplace. From the development of new, information-related products and services such as personal computers and software development; to the design of new, non-information related products such as automobiles, stereo systems and razor blades; to the construction of prototypes; to the invention of new products; to the "smartening" of everything from sewing machines to elevators to shopping carts to machine tools to buildings, the role and economic significance of information technology is rapidly expanding.

Information is, in short, taking over the world. According to Davis and Davidson, authors of *2020 Vision*, by 2020, 80% of business profits and market value will come from that part of the enterprise that is built around info-businesses. In this world of heightened info-possibility, those who know how to develop, acquire and utilize information for their companies and their customers will outpace and outsmart their competitors and win in the game of global competition. Today's business leaders unequivocally understand that when it comes to information, there are no options: either you get it, develop it and use it to your best advantage or fall into the abyss of a forgotten age.

Product Life Cycles are Shortening

In the new logistics environment, the average life span of new products is drastically declining, cycle times of everything from invoice processing to field repairs are falling sharply and the time required to complete value-added activities continues to plummet. In this environment, Japanese automakers are actively pursuing the "72-hr car;" Citicorp Mortgage is processing loans in 15 minutes; Sony Corporation is churning out a different model of its Walkman about once every three months; market and competitive intelligence data are gathered, processed and acted on in week-long cycles; and salesmen in the field are digitally connected with the production floor so that as soon as orders are placed, production begins.[7]

Throughout the developed world, the culture of change is taking root and shaping the way we think, plan, make, sell, service and structure our organizations and value-added networks. Even the very meaning of the nation-state is changing, according to former Secretary of Labor, Robert Reich, who, in his landmark book *The Work of Nations*, predicts the further pluralization and globalization of all advanced societies from America to Europe to Japan. In this new world, says Reich, "there will be no national products or technologies, no national corporations, no national industries. There will no longer be national economies, at least as we have come to understand that concept. All that will remain rooted within national borders are the people who comprise a nation."

What Reich doesn't mention is that the future he articulates will be possible largely through the creation, coordination, transmission, management and use of information and information-based technologies. It is these technologies that keep pushing the world toward one huge, interconnected economic system already foreshadowed by the rising regional economies of the Pacific Rim, the European Union, the Americas and several other regional economic alliances.

The profiles in this chapter will focus on a few of the key emerging logistics-oriented information technologies — how they are impacting individual consumers, how they are being used by companies to create competitive advantage, how, in short, they are altering the present and forming the future.

Networks, Satellites and Superhighways

Like the railroads, highways, canals and electrical and telephone systems of the industrial age, the emerging information networks promises to rewrite the rules for doing business in the information age. Networks have always been at the center of technological and civilizational progress, have always spurred drastic changes in the way goods and services are brought to market and, at a fundamental level, have changed the way business is transacted.

In 1850, on the brink of the industrial revolution, Alfred Chandler wrote about the great effort that was needed to provide and analyze the information necessary to run the complex logistics of the railroad industry. He said:

> "No other business enterprise, or for that matter few other non-business institutions, had ever required the coordination and control of so many different types of units carrying out so great a variety of tasks that demanded such close scheduling. None handled so many different types of goods or required the recording of so many different financial transactions."

Can we attempt to describe what kind of coordination and control is necessary to run today's enterprises — modern railways, airports, telecommunications networks? Since 1850, world population has more than quadrupled, the speed at which man can travel has increased from 100 miles per hour to more than 18,000 and systems — from power to transportation to telecommunications — have proliferated, grown in complexity and forever reconfigured the course of civilization.

Today, the new information networks are rewiring the planet. Although these emerging networks build on the cables and wires laid and strung during the industrial era, they are changing, improving, replacing these systems with altogether new ones that promise to hurl the world — and all that is in it — into a new reality. Known to many as cyberspace, the emerging reality is one in which previously only imagined possibilities are materializing in our midst.

Yet as much as the technology breakthroughs are giving birth to a cornucopia of novel applications and possibilities for the private user, the real potential lies in the corporate logistics sector, where businesses and organizations are capitalizing on the commercial opportunities presented by the unfolding information age. In retail, trade, manufacturing, electronics, construction — in virtually every business sector — the growing data-sphere is revolutionizing how market and customer data are gathered and used, how products and services are made and how they are distributed, sold and serviced. Some companies, such as those in telecommunications, computers and fiber optics, are transforming the marketplace while at the same time transforming themselves. MCI, for example, while building the new communications infrastructure also expects to employ that infrastructure in conducting its business. Similarly, Corning Glass Works, maker of fiber-optic cables, will benefit from the very cables it makes and sells. And Intel, maker of the much ballyhooed pentium chip, will use it to run its own personal computers.

More than any other, however, the software companies stand to cash in on the coming revolution, for it is they that will shape and define the underlying structure of the emerging logistics environment. It is their contributions that promise to be even more important and impressive than the fibers, cables and satellites through which their bits and mips race at raging speeds. It is for the brain of the information economy that the coming commercial battles will be fought, for it, more than anything, will determine the character and course of the coming logistics cyber-culture.

All these companies — MCI, Corning, Levi Strauss, Intel — and thousands more, are the pioneers of the information frontier, the engineers of the information age. They are on the vanguard of change, and it is their special contributions that will raise the entire economic machinery — from market research to product and service design to raw material extraction and processing to production, sales and after sales service — to a new level. It is their contributions that will enable companies in all sectors to design more quickly, produce more efficiently and simply do what they do better.

While there is not space enough here to cover the entire spectrum of new systems, applications, products, machines, robots, tools, gadgets and devices used in business and brought on by the information revolution, a few examples of current corporate information infrastructure developments are in order. These are the networks and mega-networks that undergird and make possible the budding info-revolution.

Local Area Networks

Local Area Networks are popping up everywhere, connecting PC users in one building or complex. From the tiny business with two or three computers to larger ones that employ hundreds, local area networks are revolutionizing the way people communicate and work together. Through electronic mail, for example, Richard Pogue, a managing partner of a global law firm, keeps communications "personal, collegiality high and far-flung offices part of the team." "E-mail," according to Pogue, "lets people converse with him who otherwise might be too intimidated to drop by his office. And while a memo may sit in his briefcase for weeks, the psychology of wanting to wipe that screen clean impels Pogue to respond almost instantly to his 35 to 40 daily electronic messages." From quarterly financial data to information about the wellness program, E-mail is changing the way people communicate within the corporation.

Bulletin Boards and Databases

Bulletin boards and data bases covering such diverse interests as new legislation, stock market performance, education for the handicapped and the weather are available for on-line perusal and interaction. With only a PC and modem, these networks allow the individual user to tap into specific sources of information and, in many cases, download that information onto a floppy or hard drive. Through a small window the size of a monitor, bulletin boards and databases enable the user to view and experience vast info-scapes and to interact with any of the hundreds of thousands of other users planet wide.

Just one such available information utility is Mintel, a system developed by France Telecom, a government-owned telephone company. With 18% of all French households hooked up to Mintel through terminals provided free-of-charge, citizens can access a variety of paid services — travel information, news, banking interest rates, and online shopping.

Internet and the World Wide Web

Of course the network of networks, the ultimate bulletin board, is Internet, an interconnected group of networks connecting academic, research, government and commercial institutions from within the United States and in more than 40 countries worldwide. In 1994, Internet linked more than 170,000 separate networks, 5,000,000 computers, and over 50 million electronic mailbox terminals, and this number continues to double each year. More than 50 million people used the mega-network, and traffic on the Internet highway continues to grow at a rate of 15% each month!

Collaborative Networks (Computer-Augmented Collaboration)

These are special electronic tools designed to support the process of collaboration so vital in the present business climate where there is a premium on team power: the process by which people unite their minds and skills to tackle problems and issues, design new products, configure new systems, write business plans. At Xerox, for example, "computer-generated shared space was found to be the best way to get people to participate playfully in meetings and a large screen becomes a community computer screen where everyone can write, draw, scribble, sketch, type or otherwise toss up symbols for community viewing. It's shared space. People can produce on it or pollute it... [The] traditional notions of conversational etiquette go out the window. One person [may write] a controversial message on the community screen while another talks about something else. Ostensibly, there may seem to be nothing revolutionary here, but exploring ideas and arguments in the context of shared space can completely transform conversation. The software injects a discipline and encourages people to create, visually and orally, a shared understanding with their colleagues. The technology motivates people to collaborate..."

Electronic Data Interchange Systems

The object here is intimacy, connectivity. The means is Electronic Data Interchange systems — electronic connections that greatly ease the burden of interacting with suppliers, customers and other entities within and external to the organization. By electronically integrating key functions, such as invoicing, scheduling and material requisitions, the value-added chain is made tighter and stronger at key linkage points. Inventories are reduced, engineering data is exchanged, work scheduling is improved, logistical distribution networks are streamlined, market feedback and research is collected and many of the former costs associated with coordination, communication and linkage are greatly reduced. More and more companies are bringing EDI systems on line. The big auto companies, for instance, now refuse to do business with suppliers who are not equipped for electronic interaction.

Expert Systems and Value-Added Networks

These systems are able to do much more than store and organize data in rigid categories, recall facts and figures and package outputs for the user. Through extensive programming, expert systems can approximate *intelligence*— in addition to storing facts, they determine and change the relationships between those facts. As new information becomes available, the knowledge base (the database of the expert system) can reorganize and repackage its outputs to reflect the new input obtained. Although still far from perfection, expert systems have saved millions of manhours and dollars in fields as diverse as medicine, manufacturing and insurance, with the application to Macrologistics not far away.

For example, Digital Equipment Corporation's XCON, one of the most successful expert systems in commercial use, has been configuring complex computer systems since 1980. The system's knowledge base consists of more than 10,000 rules describing the relationship of various computer parts. It reportedly does the work of more than 300 human experts, with fewer mistakes. Other expert systems include MYCIN, a medical system that outperforms many human experts in diagnosing diseases.

Consider the words of Alvin Toffler who writes about the human-like potential of expert systems, which he refers to as Value-Added Networks (VANs).

The existence of VANs promises to squeeze untold billions of dollars out of today's costs of logistics and distribution by slashing red tape, cutting inventory, speeding up response time. But the injection of extra-intelligence into these fast-proliferating and interlinked nets has a larger significance. It is like the sudden, blinding addition of a cerebral cortex to an organism that never had one. Combined with the automatic nervous system, it begins to give the Macrologistics System not merely self-awareness and the ability to change itself, but the ability to intervene directly in our lives, beginning first with our businesses.

Parallel Systems and Neural Networks

In the quest for speed and power, parallel processing machines use multiple processors to work on several tasks at the same time. The technology is especially promising for such applications as speech recognition, computer vision and other pattern-recognition tasks. Some supercomputers, called Connection Machines, use as many as 64,000 inexpensive processors in parallel to execute thousands of instructions simultaneously in the achievement of complicated tasks, such as those required by expert systems.

One complex parallel system is the neural network, which utilizes thousands of simple processors called neurons and which approximate the parallel structure of the human brain. Instead of processing information in linear, sequential steps according to a set of rules, neural networks are able to process information concurrently and distributively and, in doing so, learn by trial-and-error — somewhat like the way humans learn. In effect, they train themselves and form habits based on their past experience. Some neural networks, for example, are intelligent enough to narrow or widen themselves as a function of how much electronic traffic they find pulsing through their systems. That would be like Interstate 95 widening and narrowing itself automatically based on how many cars it found rolling across its surface!

Already, neural networks are being used at banks to recognize signatures on checks and at financial institutions to analyze complicated correlations between hundreds of variables and the performance of the Standard and Poor's 500 index. In the future, researchers are hopeful that the nets may provide hearing for the deaf and sight for the blind.

Corporate Virtual Workspaces

The information revolution is so important that it may end up redefining every aspect of corporate life, including the very space in which companies operate. In *Cyberspace*, authors Steve Pruitt of Texas Instruments and Tom Barrett of EDS argue that the knowledge and information society is making possible the cyberspace corporation. They write:

The traditional equation of "labor + raw materials = economic success" is rapidly changing as American businesses approach the global, highly competitive markets of the 21st century. Strategic advantage now lies in the acquisition and control of information. Corporations are becoming bewilderingly diverse and geographically far-flung. The ability to bring dispersed assets effectively to bear on a single project or opportunity is becoming increasingly difficult. The lumbering bureaucracies of this century will be replaced by fluid, interdependent groups of problem solvers.

We believe that cyberspace technology will be a primary drive toward new corporate architectures as shown in Profile 5.1. The technology will enable multidimensional, professional interaction and natural, intuitive work group formation. The technology will evolve to provide enterprises with what we call Corporate Virtual Workspaces as highly productive replacements for current work environments. Having no need for physical facilities other than the system hosting the CVW, the cyberspace corporation will exist entirely in cyberspace.

Image Processing

Image processing involves the electronic scanning and storage of paper documents, including their subsequent access. In the past, paper flows were part of the organizational functional hierarachy and division of labor. Paper became the enemy in that its management consumed many layers of workers in the typical large organization. Image processing has the capacity to reduce waste by eliminating the need to produce and store multiple copies of documents and to improve the logistics process effeciency through a single scanned image universally accessible on demand by any and all parties in the value chain.

Corporate Virtual Workspaces, websites and networks, expert systems, electronic data interchanges, image processing, Internet — these are the elements of the information infrastructure that promise to change the way we conduct business in the post-capitalist age. These are the systems that will define the structure, character and behavior of the info-corporation and become the foundation for process innovation. Add to these the throngs of information-based technological innovations — computer-integrated manufacturing, computer aided design, digital image processing, optical character and automatic speech recognition, three-dimensional modeling, virtual reality, hypermedia, digital video, computer vision, xerography and liquid crystal and plasma displays — and you've got an info-society complete with info-products from the most commonplace to the most advanced and bizarre.

Macrologistics as the Integrator

From the living room to the local ATM to the factories and offices of corporate America — and everywhere in between — the power and presence of information is linking up the planet, connecting and integrating people, companies and nations, changing and redefining the very foundation upon which modern civilization was built and upon which the logistics of industrial society was formed. Today, we rely not so much on railroads, highways, airways and shipping lanes — the logistical infrastructure of the industrial age; we rely on satellites, optical cables, computers and modems — the infrastructure of the post-industrial age. Like an expansive, convolution of interlocking spider webs girdling the globe — networks within networks within networks — the new logistical infrastructure is becoming something like the neurological system of the planet. As the integration of information technology and systems with logistics becomes real, Macrologistics Management begins to unfold in a new, more powerful way.

Macrologistics Management encompasses the whole value chain. It is an overall management framework through which to integrate the various elements of a successful channel: strategy, policy, planning, information systems, project management — all the activities required to run today's large, complex value streams. In this capacity, Macrologistics Management plays the role of integrator, focusing the entire channel in one direction, coordinating the plans and actions of hundreds of otherwise fragmented organizational sections and functions, controlling the processes and outputs all along the value-added chain and improving the business by increments and, when necessary, by leaps and bounds.

There is, then, a unique synergy between logistics and information technology: when they are properly applied together, they can catapult an organization into new levels of market strength. Macrologistics information systems work together to reduce defects, decrease cycle time, improve safety, improve delivery, increase reliability and increase customer service and satisfaction. The danger we encounter with technology is that if we do not change the way we do business at a fundamental level, it only allows us to do the *wrong things* faster.

According to Michael Hammer, the nation's foremost expert on business reengineering, technology without business process redesign is, at best, a poor allocation of resources and, at worst, a good way to automate yourself out of business. The irony is that, although we have made great and amazing advances in technology, many of us still follow business processes that were invented long before the advent of the computer, modern communications and other technological innovations.

Many business processes and methods were designed to compensate for a time when we lived in relative technological poverty. Although we are now technologically affluent, we often find it difficult to break out of the old way of thinking: that certain processes, certain methods by which business is conducted — methods that were invented decades ago — are set in stone. The practical result, according to Hammer, in the information technology industry is that many of these archaic assumptions are now deeply embedded in automated systems.

The key point about technology and logistics, then, is that both need to reflect the possibilities of an advanced technological age, not the constraints of an age gone by. When this occurs, the results can be staggering. What we learn from the profiles in this book is that the role of information technology in Macrologistics Management effort is secondary, not primary. It is a means to an end, not an end in itself. We also learn that information technology must be bridled, harnessed and controlled by the effective application of Macrologistics Management principles and methodologies. Finally, we must envision how executives and corporate leaders apply logistics to the information technology function and how information technology, in turn, supports each company's strategic intent and improvement objectives.

References and Endnotes

1. *Every Managers Guide to Business Processes,* by Peter Kewen and Ellen Knapp, Harvard Business School Press, 1996, pp. 125.
2. The Golden State University study was reported in in the Annual Conference Proceedings of the Council of Logistics Management in a paper called "*Using Information Technology to Improve Logistics Productivity,*" pp. 123. This survey recorded the 1991–1995 usage levels of information technology in logistics and the electronics manufacturing industry.
3. Woodall, Rebuck, and Voehl, *Total Quality in Information Systems and Technology,*

pp. 2–3. St. Lucie Press, 1996. We are indebted to researcher Neil DeCarlo for his contributions to the technology development protions of this chapter.
4. Golden State University study. See footnote 2 for more details.
5. Alvin Toffler, *Powershift*, New York, Bantam, 1990.
6. Alvin Toffler, *The Third Wave*, New York, Bantam, 1981.
7. James Gleick, "The Information Future: Out of Control," *The New York Times Magazine*, May 1, 1994.

Levi Strauss & Co. Takes Its Information Systems Seriously 5.1

Levi Strauss & Co. makes Levi's jeans, Dockers, and other apparel. It has manufacturing or sales operations in many countries and competes in a mature market. To accomplish its vision and mission, it has empowered its employees through autonomous teams, training and development, and information systems.

Each of the 2500 employees has a workstation and access to information in the firm's mainframe computer. Bill Eaton, the chief information officer, explains that to empower employees, it was necessary to create an open-systems architecture with complete systems access. Employees are trained in the skills needed to access various networks to obtain specific types of information. Customers and suppliers are linked to the firm through LeviLink, a network electronic data interchange system. Levi's employees even have access to their personnel files. Through OLIVER (On-Line Interactive Visual Employee Resource), an interactive computer network, they can look at up to 500 screens of personal information. Using OLIVER, employees can check their total compensation, disability, health care, pension, employee investment plan, survivor benefits, beneficiary, and other personal information.

All the information systems at Levi Strauss, from purchasing to personnel, can communicate with each other. In fact, the capability for such communication is considered in all of the company's technology purchases. Levi Strauss has also created a special position, Director of Quick Response, to deal with electronic services to retailers and suppliers. Expert systems are used in several areas, such as inventory management. Finally, the globalization of its business has caused the firm to install international communication networks that enable its far-flung operations to keep in touch with each other.

 Sources: David Brousell, "Levi Strauss's CIO on: The Technology of Empowerment," *Datamation* (June 1, 1992), pp. 120–124; Jennifer J. Laabs, "OLIVER: A Twist on Communication," *Personnel Journal* (September 1991), pp. 79–82; and Larry Stevens, "Systems Development vs. the Tower of Babel," *Bobbin* (June 1992), pp. 22–24.

How Architecture Wins the Technology Wars

5.2

The global computer industry is undergoing a radical transformation. Success today flows to the company that establishes proprietary architectural control over a broad, fast-moving, competitive environment. Since no single vendor can keep pace with the outpouring of cheap, powerful, mass-produced components, customers have been sewing their own patchwork quilt of local systems solutions.

The architectures in open systems impose an order on the system and make interconnections possible. It is the architectural controller who has power over the standard by which the entire information package is assembled. Microsoft Windows popularity is used as an example to show how companies like Lotus must conform their software to its parameters in order to have market share. Thus, the concept of proprietary architectural control has broader implications in that architectural competition is giving rise to a new form of business organization. The experts contend that a small handful of innovative companies will define and control a network's critical design.

It is necessary to cannibalize old niches in order to evolve to occupy an ever-broader competitive space. The Silicon Valley model is used to show four important operational features that underlay the overall basic theme of the article: inventing — and reinventing — the proprietary architectures for open systems is critical to competitive success and can serve as the platform for a radiating and long-lived product family.

Overall, the five basic imperatives that drive most architectural contests are:

1. Good products are not enough.
2. Implementation matters.
3. Successful architectures are proprietary.
4. Open general-purpose architectures absorb special purpose solutions.
5. Low-end systems swallow high end systems.

Of added value in this story are three sidebars featuring scenarios for architectural competition:

1. Graphical user interfaces.
2. Video games.
3. Page-and-image description standards.

The story ends with a look at Xerox's failure to capitalize on its pioneer xerography niche and not creating spin-off industries and business lines. The authors end with the challenge: "We think that similar strategies are available to companies in other complex industries — aerospace and machine tools, among others. If so, the information (technology) sector's strategic and organizational innovations might prove as interesting as its technology." One of the better HBR articles to come along in the Information Technology field in the 1990s.

Source: **Charles R. Morris and Charles H. Ferguson. *Harvard Business Review (HBR)*, March-April 1993, pg. 86–96.**

Achieving Logistics Quality Through Intelligence 5.3

Information — in any form — can empower employees to make continuous improvements in product and process, by monitoring customers, competitors, and suppliers. This profile describes the complexities of information sharing at Corning, where 25,000 employees in 90 locations throughout the world must communicate about 60,000 products and services. A "Corrective Action Team" was formed to conduct an information audit. This resulted in a computer-based "Information Exchange Intelligence System" that allows information entry and access throughout the company. Cost per user is lowered by making the system accessible to all employees, and instead of funneling all information through the company's information systems department, data can be entered by any system user in a standard format. By creating a personal electronic file folder, any time new information that matches a user's request is fed into the system, it is entered into the user's folder for viewing the next business day. A sidebar to the story — based on a survey of over 200 large companies highlights six principles for building an information system for logistics quality. The bulk of the article discusses the way information flow is important to the five key factors assessed by Baldrige Award examiners:

1. Customer focus
2. Meeting commitments
3. Process management and elimination of waste
4. Employee involvement and empowerment
5. Continuous improvement.

A related article in *Long-Range Planning*, 2/92, gives some examples from a number of companies, such as Xerox, M&M Mars, Johnson & Johnson, IBM, GTE, and Milliken.

Source: Fuld, Leonard M., *Long Range Planning*, February 1992, pp. 109–115.

Kao Corporation's Information System Supports its Innovation Strategy 5.4

Japan's Kao Corporation, begun as the Kao Soap Company in 1890, is a diversified company with interests in two primary areas: household products (with divisions for personal care cosmetics, laundry and cleansing, and hygiene) and chemical products (with divisions for fatty and specialty chemicals, and floppy disks). The firm was founded on the principle of equality as expressed by 7th-century statesman Shotoku, whose philosophy profoundly influenced Yoshio Maruta, Kao's CEO from the late 1970s through the early 1990s. Two of Shotoku's precepts are: "human beings can live only by the Universal Truth, and in their dignity of living all are absolutely equal," and "If everyone discusses on an equal footing, there is nothing that cannot be resolved." Kao's corporate philosophy thus reflects belief in individual initiative and rejection of authoritarianism.

Because work is viewed as something flexible and flowing, Kao is designed to be run as a "flowing system" that spreads ideas and stimulates interaction. To give free rein to creativity and initiative, organizational boundaries and titles have been abolished. The result is a learning organization in which sharing of information is essential and information systems are critical to the company's success.

Every employee at Kao seeks to learn and help others learn. Every manager knows Maruta's fundamental assumption: "In today's business world, information is the only source of competitive advantage. The company that develops a monopoly on information, and has the ability to learn from it continuously, is the company that will win, irrespective of its business." In some U.S. firms, executive information systems allow only top managers to gain access to key data; at Kao, the MIS extends to everyone. The equality perspective makes information available to everyone in the company. The task of Kao's managers is to take information from the environment, process it, and by adding value, turn it into knowledge. Every piece of information is viewed as potentially providing insight into product positioning, product improvement, or the development of a new product.

Nowhere is the advantage of this philosophy more evident than in Kao's flexible manufacturing program designed, according to systems developer Masayuki Abe, "to maximize the flexibility of the whole company's response to demand." The firm obsessively collects and distributes data, which are entered into a single system that links together sales and shipping, production and purchasing, accounting, R&D, marketing, hundreds of shopkeepers' cash registers, and thousands of salesmens' hand-held computers. Kao boasts that its management information system is so complete that it can turn out an annual report one day after the end of the year.

Kao can tell whether a new product will be successful within two weeks of the product launch. It does so by using focus groups, consumer calls, and point-of-sale information from 216 outlets in a system known as Project Echo. This approach obviates the need for market surveys. It also helps explain how Kao could enter the highly competitive cosmetics industry in Japan and become the number two player in that industry in less than ten years — to put it succinctly, the company can adjust quickly to meet customer demands. When, for example, Mrs. Wanatabe buys a bar of soap, the purchase is instantly recorded. With this information Kao can increase variety while cutting inventory levels. In addition, through its wholesalers Kao can deliver an order within 24 hours to any of 280,000 shops that, on average, order only seven items at a time. Perhaps no other company in the world can match Kao's flexibility.

© James M. Higgins. *Sources:* Thomas A. Stewart, "Brace for Japan's Hot New Strategy," *Fortune* (September 21, 1992), pp. 62–73; and Sumantra Ghoshal and Charlotte Butler, "The Kao Corporation: A Case Study," *European Management Journal* (June 1992), pp. 179–191.

Better Information Means Better Quality 5.5

What role can information systems play in achieving logistics quality? This story examines 3 main areas where information technology can play a very important role:

1. **Process monitoring.** This takes many forms: collecting performance data, modeling possible scenarios, or tracking the results of quality efforts. By monitoring a process closely, it is possible to detect and fix problems early.
2. **Customer Service.** The author gives the example of a software company whose goal was "to make it as easy as possible for the customer to do business with it and to work with its product." A 50% share of a very competitive market was the result of their efforts. Quality in customer

service refers to every interaction that might take place between the customer and the company, from the moment of purchase, to delivery, maintenance, etc. Quality is solving your customer's problems. "Information systems," says Michael Ashmore, "can play a critical role in providing answers quickly and accurately, in every aspect of the service strategy."

3. **Production.** Production systems can contribute to logistics quality in two ways: (a) production support, which includes yield, productivity, and cost control; and (b) cross-functional information exchange between production and other basic business functions within the extended enterprise Value Chain.

Information systems can have a great impact in making logistics quality a reality. And if one applies the total quality philosophy to the information systems themselves, then you can increase their capacity to be more responsive to the needs of "customers (both internal and external) and the quality needs of the business."

Source: G. Michael Ashmore, *The Journal of Business Strategy*, January-February 1992, p. 57–60.

Is Data Scatter Subverting Your Strategy? 5.6

The main premise of this article is that many organizations have information available from many sources to let them know how they are doing. The problem is the poor coordination of this data and its frequently nebulous connection with the organization's purpose. The solution lies in systems and processes for overcoming the data scatter.

When viewed from the perspective of information systems, most organizations are Balkanized environments: bits and pieces scattered in often conflicting jurisdictions such as MIS, marketing, finance, sales, HR and so forth. They have a 21st century computing capacity and 1960s approach to transforming data into useful information. Data scatter results from the poor job that organizations do in managing and transforming data into integrated information needed to drive success within the Value Chain.

The symptoms of data scatter that should alert you to possible trouble are:

- **Strategy Silos:** where each executive defines the future in terms of their own self-image which is sub-optimized to the overall detriment of the Value Chain.

- **Data Wars**: where knowledge is power and weapons pointed at colleagues often start a family feud. Bootleg data bases abound.
- **Decision Jerk**: occurs when management zigs and zags with each piece of data, continually shifting priorities or piling on new initiatives. Chunks of data are used to make decisions resulting in multiple misguided efforts.

Lingle and Schiemann offer **Five Measures of Success** with the key toward focus on the Value Chain:

1. Strategically anchor gauges
2. Reflect the outcomes — not the activities
3. Create a counterbalance
4. Ensure responsiveness to change
5. Exhibit strong signal-to-noise characteristics.

They point out that winning organizations focus their measures around a balanced scorecard type of system in a family of measures approach. To maintain focus, it is important that the organization track on a limited number of gauges, which are updated regularly and available to all the people in the workforce. To the extent they can avoid data scatter, they will be able to tap into that storehouse of information and convert it to new knowledge for serving customers in a more innovative way than the competition.

Source: John H. Lingle and William A. Schiemann, *Management Review (MR)*, May 1994.

Livonia Recreates Its Quality Information Systems 5.7

This thought provoking article discusses the important role of information systems in implementing logistics-driven Total Quality systems and explains how one company, Livonia, revamped its management information system to create "quality information systems." With annual revenues over $700 million and a workforce of 4,500, Livonia enjoyed gains in product quality and cost reduction, but remained stagnant in the "office environment." Specifically, the general business processes were being maintained in the status quo while the Total Quality focus was on the plant floor.

The article discusses management's realization that MIS occupy a central focus of the company's TQ movement and must be combined in a strategic way in order to achieve a new understanding of system and service. This involves expanding the scope of the "data processors" to include the entire business process, including those that have little to do with the IS function or department. QIS requires the systems department to develop, implement, and champion a methodology that looks at the business goals and develops activities that might or might not include a computer system to meet those goals. Service should be the MIS department's deliverable product. This service should include setting goals, facilitating and participating in process improvements and implementing computer systems.

The MIS function took over the business reengineering activities and a sister group to "applications development" was formed with senior business analysts from technical and non-technical backgrounds. Three goals were formed:

1. Everyone must understand the QIS initiative.
2. MIS/QIS is uniquely suited to facilitate process improvement.
3. The goals of the company must be the same as the goals of MIS/QIS.

The group found success using the process enhancement techniques described in *Improving Performance,* by Rummler and Brache (Jossey-Bass, San Francisco, 1991). Flowcharts were developed to depict the customer/server relationships and the corresponding processes being performed. These flowcharts were based upon performance metrics and standards that focused on key processes of providing price quotes to customers, developing installation drawings, and entering orders. None of the standards set by the team were met once an order was received. The marketing division was reorganized and work cells were created having total responsibility for any job assigned to its area from start to finish. Each cell was measured and held responsible for errors, quality problems, and substandard performance. Supervisors were eliminated resulting in better information flow between the company and the customer.

Also important are the long term considerations for procedural improvement focused in two areas: education of the customer and automation of the remaining streamlined processes. One of the most important aspects of the overall effort was the buy-in by division management and staff. Six months after implementation, the prime objectives were being met in all areas leading to the conclusion that the combination of QIS and an improved bottom line creates a win-win situation for all with the Value Chain.

 Source: MIS + TQM = QIS, Richard Keith, *Quality Progress (QP)*, April 1994.

ELM: A Holistic Approach 5.8

The author of ELM is a well-known business writer who presents a concise look at one of the key features of how to use Information Technology to more successfully coordinate logistics throughout the organization in order to more reliably deliver goods and services to customers. ELM stands for Enterprise-Logistics Management, which is an important new step in the passage of manufacturing from art form to science.

ELM often quotes author Thomas Gunn, using excerpts from his book *Age of the Real-Time Enterprise: Managing for Winning Business Performance Through Enterprise Logistics Management* (Oliver Wright Publications). Gunn defines ELM as a holistic approach to managing operations and the value-added pipeline (total supply chain), from suppliers to end use customers.

John Teresko, logistics expert discusses three drivers to achieve competitive position: (1) a relentless quest for customer satisfaction, (2) recognition of the need for real-time management, and (3) ability to perform in a world-class manner. Additionally, superior logistics management is increasingly being cited as the new strength of the Japanese manufacturers, instead of JIT alone. The post-JIT environment places an emphasis on information systems and technology, including electronic production-control systems. According to Prof. Jichiro Nakane of the Systems Science Institute, Tokyo, the key is to achieve superior management of the flow of information, from customer order to the delivered product or service, with an integrated enterprise-wide system.

The shortcomings in today's software solutions are discussed in terms of MRP products/vendors inability to execute DRP (distribution resource planning) at the front end and a lack of procurement applications at the back end. The inability to track material through the entire manufacturing process results in poorly tracked plans and schedules. When computing is diffused throughout the entire organization, it is difficult to collect and understand cost allocation trends attributed to open, distributed architecture. Teresko concludes with the premise: there is ample evidence that the total cost of spending on information systems in a typical company may represent 30 to 40% of its capital spending. And they may be surprised to learn that in their catch-up mode, they are not spending nearly enough.

Source: John Teresko, *Industry Week (IW)*, June 20, 1994. Penton Publishing, Cleveland, Ohio.

MOBILIZATION

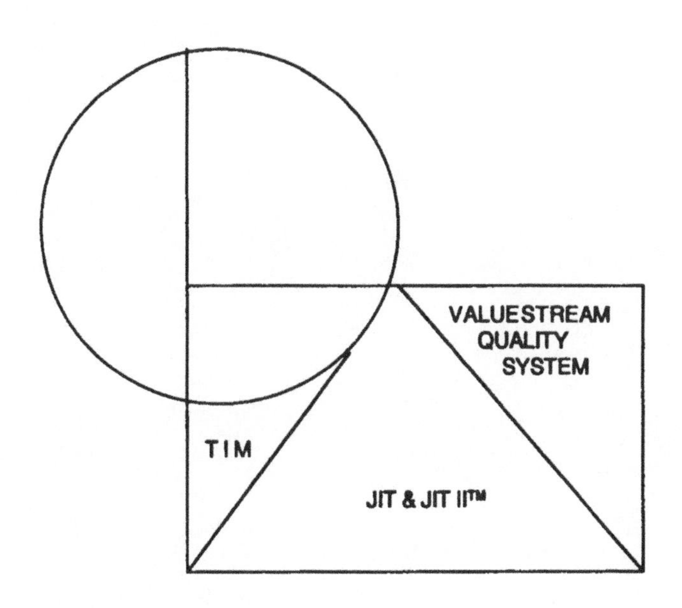

6 Total Innovative Management: The Integration of Innovation, Management Quality, and Adaptive Learning

During the past 25 years or so, the logistics management model has evolved in three phases, each covering a 10-year timeframe. The first phase was the restructuring phase of the late 1960s and most of the 1970s. It consisted of a series of loosely connected crafts, with random applications of cost cutting techniques focusing on the cost of each activity. The overall strategy was one of negotiation, substitution and productivity squeezes.

Computer power is now 8,000 times less expensive than it was 30 years ago. If we had similar progress in automotive logistics technology, you could buy a Lexus for about $2 today. It would travel at the speed of sound and go about 600 miles on a thimble of gas."

John Naisbitt

The second phase took place during the 1980s and focused on defining, understanding and mobilizing the supply chain. As a result, two or more crafts pulled together and there was improvement to the entire supply chain, bringing customers and suppliers into the supply system. The objective of these logistics actions was to remove unnecessary steps and excel in customer service, while achieving dramatic cost reduction and elimination.

The third phase, which began in the early to mid-1990s, can be described as business transformation, with Macrologistics as the major thrust forward for mobilization. Its core concept now revolves around the principles of competitive advantage in logistics, when properly deployed can expand market share dramatically.

Recently, Myron Tribus noted that Macrologistics Management begins with the re-definition of leadership, as adherents to Quality Management have also discovered:

> "One of the most frequent reasons for failed logistics, as well as Quality Management efforts, is that many leaders and managers are unable to carry out their responsibilities because they have not been trained in how to improve the Management Cycle, which is the key one of twelve elements in the concept of Total Innovative Management.
>
> Because managers do not have a well-defined template and process to follow— a process founded on the principles of customer satisfaction, respect for people, continuous improvement and speaking with facts, they often fail. Dr. Deming, who is widely regarded as the pioneering spirit of the Quality Management movement, taught the importance of focusing on the following ten leadership actions for improving the Macrologistics Management Cycle component of Total Innovative Management (TIM):[1]

1. Recognize the Management Cycle as a system.
2. Define it so others can also recognize it.
3. Analyze its behavior.
4. Work with subordinates in improving the system.
5. Measure the quality of the Management system.
6. Develop improvements in the quality of the system.
7. Measure the gains in quality, if any, and link these to customer delight and quality improvement.
8. Take steps to guarantee holding the gains.

9. Attempt to replicate the improvements into other areas of the system.
10. Tell others about the lessons learned."

The TIM model depicts the coming together of some of these concepts in a logical fashion. One of the core concepts underlying Total Innovative Management is that of a Systems view. Specifically, from the design point of view, TIM is really three systems in one: (1) the freight system, (2) the warehousing system, and (3) the inventory management system. Supporting the three systems are the four principles of Total Quality Management: (1) Stakeholder Satisfaction, (2) Continuous Improvement, (3) Speak with Facts, and (4) Respect for People. These four principles are interrelated, with the Stakeholder Satisfaction Principle at the core or the hub of the Management Quality system. It is the Management Cycle which heavily influences the logistics system throughout the value chain.

Leadership and Total Innovative Management

Total Innovative Management (TIM) is total in three senses: it covers every process, every job, and every person in the value chain. First, it covers every process, not just manufacturing or production. Design, construction, R&D, accounting, marketing, repair, and every other function must also be involved in quality improvement. Second, TIM is total in that every job is covered, not just those involved in making the product. Secretaries are expected not to make typing errors; accountants not to make posting errors; presidents not to make strategic errors. Third, TIM recognizes that each person is responsible for the result of his or her work and for the work of the group.

TIM also goes beyond the traditional idea of quality which has been expressed as the degree of conformance to a standard, or the product of workmanship. Enlightened organizations accept and apply the concept that quality is "the degree of user's satisfaction, or the product's fitness for use." *In other words, the stakeholder determines whether or not quality has been achieved in its totality.* Accordingly, we learned that for mobilization to occur, Macrologistics assumes that TQM is operational using a TIM framework, and that the organization has committed itself to the principles of Adaptive Learning. It is then that Macrologistics Management is used to create strategic advantage by leveraging the unexploited potential of logistics to mobilize the organization to become more customer focused and to truly practice continuous improvement.

The same measure — *total stakeholder satisfaction* — applies throughout the entire management cycle of the organization. Only the leadership team and the "outer edges" of the company actually have contact with customers in the traditional sense, but every other department can treat the next department as its customer. The main judge of the quality of work is the stakeholder, and if the customer isn't satisfied, the work doesn't have quality.

In that regard, it's important, as the Japanese say, to *"talk with facts and data."* TIM emphasizes the use of fact-oriented discussions and statistical quality control techniques by everyone in the value chain, not just engineers or management. TIM insures that everyone in the enterprise is exposed to basic quality control ideas and techniques and is expected to use them. Thus, TIM becomes a common language and improves "objective" communications between the leaders and their people.

However, TIM is more than the attempt to make better products; it's also a search for value-added ways to make them. Adopting the TIM philosophy commits the extended enterprise to the belief that there's always a better way of doing things, a way to make better use of the company's resources, a way to be more productive. In this sense, TIM relies heavily upon value analysis and value engineering as a method of developing better products and operations in order to maximize value to the stakeholder — customers, employees and shareholders.

Total Innovative Management (TIM) Underlying Concepts

First and foremost, Total Innovative Management is a set of philosophies and management systems directed to the efficient achievement of each organization's objectives in the value chain to assure customer satisfaction and maximize stakeholder value. This is accomplished through the continuous improvement and mobilization of the Logistics Quality System consisting of the Freight System, the Warehouse System and the Inventory Management System. Thus, it becomes a way of life for doing business for the entire organization.

It is a concept that says a company should design quality into its products, not inspect for it afterwards. Only by having a devotion to quality throughout the company will the best possible products be made. TIM is also too important to take second place to anything else in the company's goals. Most specifically, it should not be subsidiary to profit or productivity. Concentrating on quality will ultimately build and improve both profitability and productivity. Failure to concentrate on it will quickly erode profits as "customers" resent having to pay for products they think are low quality.

Total Innovative Management's main focus is on "why." It goes beyond the "how-to" to include the "why-to." It tries to find the causes of defects throughout the supply chain in order to remove them. It's a ceaseless round of finding defects, finding their causes, and improving the process to totally eliminate the causes of defects. Accepting the idea that "the next process in the extended enterprise is our customer" is essential to the real practice of TIM which says that each process should not only develop its own process control charts but should also disclose its own errors to the next process in order to raise quality. However, it has been said that it seems contrary to human nature to seek out one's mistakes. We tend to always find the errors caused by others and to neglect our own. Unfortunately, that self-disclosure is what's really needed.[2]

Too often instead, management tends to blame and then take punitive action. This attitude prevails from first-line supervisors all the way up to top management. In effect, we are encouraged to hide the real problems we cause, and instead of looking for the real causes of problems, as required by Total Quality, we look the other way.

TIM and The Notion of Control

The Japanese notion of "control" differs even more radically from ours; that difference of meaning does much to explain U.S. management's failure to adopt Total Quality in the 1980s and the 1990s. "Control" carries, for us, the meaning of someone or something limiting an operation, process, or person. It has the overtones of a "police force" in the industrial engineering setting and is often resented as a Gestapo-type function. In Japan, as pointed by JUSE counselor and Japanese QC scholar, Dr. Noriaki Kano, *control* means "all necessary activities for achieving objectives in the long-term, efficiently and economically. Control, therefore, is doing whatever is needed to accomplish what we want to do as an organization."

Ray Stata, CEO of Analog Devices, observes that Japan was the first nation to achieve economic success based upon management quality and innovation. Virtually all previous breakthroughs in national economic success had been based upon technology. But Japanese organizations brought new ways of managing to the same essential product and process areas as their North

American counterparts and used those improved management quality techniques to transform Japan into a major world economic force.[3]

Total Innovative Management as a System

As outlined by Sericho Yahagi, there are twelve factors and some forty-one sub-factors involved in the mobilization system of Total Innovative Management. These factors, along with their associated sub-factors, are as follows:

- **Corporate History:** past, present and future.
- **Corporate Climate:** core, climate and culture.
- **Strategic Alliances:** objectives and coherence.
- **Channels:** suppliers and buyers.
- **Management Cycle:** vision, strategy, planning, organizing, implementing and controlling.
- **Environment:** economic, societal, and global.
- **TIM Targets:** inputs, markets, technologies, and products.
- **Business Structure:** business fields, business mix, and market standing.
- **Management Resources:** money. materials, information, and people.
- **Management Design:** system, organization, authority, and responsibility.
- **Management Functions:** decision-making, interrelationships, and quality.
- **Measurement System Performance:** growth, scale, stability, profit, and market share.

Of the twelve, six are critical to the success of the company's Macrologistics Management System: The Management Cycle, Business Structure, Management Resources, Management Design, Corporate Culture, and Measurement System Performance. These six factors are interrelated in a circulatory system of factors which have a cause-and-effect relationship as follows: The Management Cycle acts as the driver and influencer of the four factors of structure, resources, design and culture — which in turn affect measurement system performance.

If the quality level of the management cycle is low, then the six factors generate a negative or bad feedback loop which finds management passively waiting until poor results of management performance forces reactionary feedback into the management flow.[4] If the quality level of the management cycle is "high", then these six factors generate an excellent feed-forward loop in which management perceives the strategies and plans needed for success of each management factor and then proactively formulates and implements them.

In other words, a "bad" or poor management cycle drains the quality from the other management factors and an excellent management cycle pumps up the quality of the other four factors. Thus, the synthesized effects of these four factors cause the "score" of the measurement system performance to be either higher or lower than the management cycle, depending upon the good or bad conditions present. Finally, because management performance flows into the management cycle, the cyclical corporate growth of the "best score" companies will incrementally improve the management cycle and ultimately overall management quality.

As can be seen in the studies by Yahagi, comparing the scores of the "best" and "worst" companies, the Management Cycle score clearly drives the final effect on the Management Performance score. The visible differences between these organizations was very clear: proactive vs. passive; perceptive vs. reactionary; energizing vs. draining; and feed-forward vs. feedback..[5]

Of particular interest and importance are the responses for the first two elements in the Management Cycle — corporate vision and management strategy. In both cases, the highest number of respondents scored the maximum of 5 points for both categories. In the case of the vision, there is a distinctive organizational corporate vision, which is stipulated as a long-term management direction. In the area of strategy, a comprehensive Management Strategy exists which has been created from the vision and has been linked to planning.

Application to the Three Logistics Systems

The inbound and outbound freight expenses are typically the largest logistical outlay for the business enterprise. The time-honored cost-reducing process that has delivered over and over is to focus on a few carriers who consistently meet the delivery requirements. If you have a private fleet, how well you are using the equipment is a key focus point. And don't overlook the possibility of picking up raw materials in order to save money and time.

Warehousing is typically the second largest expense area and its management system tends to be very labor intensive, so whatever can be done to make labor more efficient reduces costs. Finally, the inventory management system needs to be addressed, as discussed in the next chapter on JIT and JIT II™.

Innovation as the Key to Logistics Productivity

The United States has experienced sharp setbacks in numerous industries in recent years. We have witnessed the competitive position of U.S. firms decline in autos, steel, textiles, appliances, cameras, computer chips, and others. Peter Drucker once pointed out that the one core competence needed by every prosperous organization is innovation.[6] *Fortune* magazine writer Brian Dumaine, after examining these trends in various businesses, observed that American businesses are not creating products fast enough for global competition.[7] U.S. companies are facing a more competitive international environment than ever before and need to take action to halt these trends, as some organizations have already successfully done.

The 3M company is famous for its never ending renewal process of producing new products, which have included Scotch Brand cellophane tape and the international favorite, Post-It notepads. Bell Labs has consistently produced a large number of successful new products, among them the transistor and fiber optics. And almost every year, year in and year out, General Electric (GE) files for more U.S. patents than any other company headquartered in the U.S.[8]

Ford Motor Company continues to cut production costs, improve product quality, and utilize new management systems that increase productivity.[9] Meanwhile, Toyota's constant cycle of innovations may someday make it the #1 automobile company in the world, profiting handsomely from the Lexus and from the relative low-cost advantage that no other firm can match.[10] Chrysler continues to amaze the auto world with new products like the Concorde, LHS and Neon, the cab-forward concept, all of which are sold at reasonable prices.[11] Xerox cuts costs, brings out new products at a dazzling rate, and revolutionizes its approach to management. Hewlett-Packard continues to steamroll the competition and grow at a staggering pace, in spite of its $30 billion, 100,000 employee size, launching new products at a rate that few companies can match. Many point to its innovative ways as a major reason for its success.[12]

What these firms have in common is a remarkable proclivity for innovation, both in products and services. In fact, virtually all leading authorities on business, including Fortune 500 CEOs, researchers, and consultants agree that there is only one way firms can cope with all the challenges confronting them in the 21st century to come is to innovate. They have all recognized that in today's environment, they must either **innovate** or **evaporate.**[13]

Innovation and TIM

Innovation is the process of creating something new that has significant value to an individual, group, organization, industry, or society. In other words, an innovation is a creation that has significant value. Innovation is how a firm makes money from its creativity. To do so, a firm needs to distinguish between ideas as being original or creative. Original ideas just aren't enough. The ideas generated must have the potential for significant value, thereby becoming innovations. Organizations will not be as effective or efficient as they should be if they cannot innovate. Solving problems and pursuing opportunities requires solutions, many of which are unique to the specific situation.[14]

However, before we can have innovation, we must have creativity, which is the skill to originate something new and to make it valuable. Everyone possesses some form of innate capacity for creativity. But in most individuals, the transformation of the capacity into a skill has been stifled by parents, teachers, and bosses who enforce the rules about what is acceptable behavior. Because only a few behaviors are allowed, the creativity of exploring new realms is stifled and must be unleashed to tap full potential.[15]

The Four Types of Innovation

According to innovation experts , such as James Higgins and Paul Plsek, there are basically four types of innovation that organizations involved in Macrologistics Management strategies need to be concerned about: product, process, marketing, and management.[16] These four types are coupled with some forty characteristics that an organization's culture needs to possess to achieve strategic competitive advantage through innovation and adaptive learning. Higgins uses four questionnaires that an organization can use to measure its performance in this area.

Product innovation results in new products or services, along with enhancements to existing ones. Toyota and Chrysler are good examples of product innovation. Process innovation results in improved processes within the organization's supply chain, thereby focusing on improving the efficiency and effectiveness of logistics processes. Toyota and Cooper Tire are good examples of logistics process innovation. Marketing innovation is related to the marketing functions of promotion, price, distribution, and product-related packaging. For example, in advertising its new products, Chrysler uses innovative marketing themes such as those featuring its $1 billion Technology Center. Finally, management innovation improves the way the organization is managed. Bell Labs uses total innovative management strategies, such as programs aimed at improving researcher's productivity, to create new products.[17]

Product innovation leads principally to competitive advantage through differentiation, while process innovation leads to a low-cost advantage. In "World-Class" organizations, product and process innovation are coordinated. A 15-year study by Varajaradan and Ramanujam, performed on successful companies, found commitment to product and process innovations to be one of the six common Critical Success Factors of these firms.[18] Most of the successful firms in the U.S. and Japan, and to some degree in Europe, integrate process and product integration.[19]

Marketing innovation helps achieve relative differentiation and relative low-cost objectives by improving strategies and tactics concerned with the marketing mix. In an age where the consumer is bombarded by advertisements, innovative marketing techniques are crucial to successful sales. In many cases, real differentiation and lower cost don't matter if the customer perceptions are not in line. Marketing innovation helps align the desired perceptions by building strong relationships with customers so that the five key marketing mix variables are strengthened. These marketing mix variables are product, promotion, price, distribution, and target market. See the Profile 5.1 (IKEA) for more details.

Management innovation can help achieve both differentiation and low-cost competitive advantage by improving the efficiency and effectiveness of logistical efforts to achieve corporate goals. U.S. and Canadian firms must improve their management practices to compete successfully in the global marketplace. As logistics pioneers have often pointed out, the manager's primary function is creative problem solving in the areas of the management cycle: planning, organizing, leading and controlling. As organizations continue to move toward greater self-management, individuals at all levels will have increased responsibility for creative problem solving.

Higgins offers the following examples of Management Innovation:

- **Creative Problem Solving:** expert systems, lateral thinking knowledge management, adaptive engineering.
- **Planning:** strategic alliances, joint ventures, scenario forecasting, business plan software, speed strategies.
- **Organizing:** reengineering, process redesign, creativity circles, restructuring, intrapreneurship, networked organizations.
- **Leading:** transformational leadership, empowerment, whole systems change conferencing, management by wandering around.
- **Controlling:** Comshares "Commander" Executive Information System, self-management, activity based costing, interactive balanced scorecard measurement systems.

Summary

Strategic challenges will force management to change and innovate in the next decade, sometimes dramatically and sometimes incrementally. Although many of the changes have already begun, as shown in the profiles contained in this book, many more will be needed. Total Innovative Management can function as a change accelerator by developing new ways of managing change and stress, as well as improved information systems. Increasing the competition factor will demand improvements in competitor marketing intelligence systems, as well as new strategies for customer linkages.

The globalization of business will require logistical global strategies and structures, global cultures, and creative management styles. Changing technology will provide a new way of accelerating product life cycles and new product concepts to achieve new competitive advantages. A diverse workforce will require new leadership and management styles, new benefits, and new reward systems. The transition to a knowledge based society will require new, innovative management paradigms and knowledge management techniques. Finally, increasing complexity will demand more expert systems and computer simulations, which will give Total Innovative Management a prominent place in the Macrologistic System of managing logistics resources and processes.

References and Endnotes

1. *The Systems of Total Quality,* Myron Tribus, Self-published, 1990.
2. *A History of Science,* C. D. Whethan, 4th Ed., McMillan & Co., 1980.

3. Ray Stata, "Organizational Learning: The Sustainable Competitive Advantage", *Sloan Management Review*, Spring, 1989, pp. 63–74.
4. According to Yahagi, the circulatory aspects of the management quality system are a key to understanding the relationships of this system to others, such as the Malcom Baldrige National Quality Award.
5. For further details, see the source article 'After Product Quality in Japan: Management Quality', by Seiichiro Yahagi (*National Productivity Review*, Autumn 1992).
6. See "The Information Executives Truly Need", Peter Drucker, *Harvard Business Review*, 1994.
7. Brian Dunaine, "Closing the Innovation Gap", *Fortune*, December 2, 1991, pp. 57–59.
8. Peter Coy, "The Global Patent Race Picks Up Speed", *Business Week*, August, 1994, pp. 57–58.
9. Raymond Serafin, "Troutman's Tenure May See Ford Sieze the Lead". *Advertising Age* (October 11, 1993), pp. 1, 50; and Alex Taylor III, "Ford's Million Dollar Baby", *Fortune* (June 28, 1993), pp. 76–81.
10. James Higgins, "*Innovate or Evaporate*", New Management Publishing, Winter Park, Fl, 1995, pp. 4; William Spindle, " Toyota Retooled", *Business Week*, April 4, 1994, pp. 54–57; "Toyota Puts It on the Line", *U.S. News and World Report*, August 23, 1993, pp. 47–48.
11. David Woodruff, "Chrysler's Neon", Business Week, May 3, 1993, pp. 116–126; "Eaton's Plan: Keep Focused", *USA Today*, May 10, 1993, pp. 38.
12. James Higgins, *Innovate or Evaporate*, New Management Publishing, Winter Park, Fl, 1995, pp. 4.
13. Ibid., pp.8.
14. Ibid., pp. 10.
15. Ibid., pp. 16.
16. Ray Statta, "Organizational Learning — the Key to Management Innovation", *Sloan Management Review*, Spring 1989, pp. 63–74; also see James Higgins' seminal work on innovation called "*Innovate or Evaporate*", in which he describes in detail the four types of organizational innovation, along with a description of the culture change required to foster innovation among all employees. See also *Creativity, Innovation, and Quality* by Paul Plsek, ASQC Quality Press, 1997.
17. Ibid., pp. 24; also see Robert Kelley and Janet Kaplan, "How Bell Labs Creates Star Performers", *Harvard Business Review*, July–August, 1993, pp. 128–139.
18. P. Rajan Varadarajan and Vasudevan Ramanujam, "The Corporate Performance Conundrum: A Synthesis of Contemporary Views and Extention", *Journal of Management Studies*, September, 1990, pp. 463–4819.James Higgins, "*Innovate or Evaporate*", New Management Publishing, Winter Park, Fl, 1995, pp. 55. Higgins also presents an integrated approach which focuses on process innovation, a Macrologistics driver. In this seminal work, he has focssed on seven process innovation areas: marketing, operations, finance, HR management, information systems, and technology, research and development, and management. Plsek takes a similar approach in his work cited in Endnote 16. Also see the studies of Masaaki Kotabe published in the *Journal of Marketing*: "Corporate Product Policy and Innovative Behavior of European and Japanese Multinationals", April, 1990, pp. 19–33.

Burlington Northern Achieves Breakthrough: The Little Engine That Could 6.1

Mission/Vision

Our mission as a company is to provide the many markets we serve with products of consistently superior quality at price levels that are fair and competitive. Achieving this mission is a responsibility that we all share and is necessary to meet the expectations of our customers, ourselves, and our shareholders. With this uncompromising dedication to superior quality, we have a focus for our actions that unifies us, adds value to our work, and enriches our lives.

Macrologistics Strategy: use of small-scale teams to achieve breakthrough results which are tied to performance goals and key business issues. The focus is on improving the macro-process of the extended enterprise. This involved the key concept of installing Breakthrough Thinking throughout the entire value chain using a combination of cross-organizational project teams, EDI based Information Systems, and Process Enhancement techniques. The organization has achieved tens of millions of dollars in 'savings' while reducing the cycle time for delivery by using over 200 Breakthrough teams.

Background Information

In 1992, the Chief Operating Officer and President of Burlington Northern Railroad, Bill Greenwood, received a suggestion from a friend that would transform his thinking and that of his company as well.

> "My friend suggested that I read a book by Robert Schaffer called 'The Breakthrough Strategy'. It was written by a consultant from Connecticut who went on to popularize the concept through a couple of articles in the *Harvard Business Review.* Well. my friend thought there may be some possible applications of the ideas at BN and it turns out he was correct. It turned out to be a million dollar idea".

Over the next few years (1992 to 1993), Burlington Railroad would begin to put together its Macrologistics strategy around the four following components:

- Small-scale teams
- Urgent, measurable goals
- Real business issues involving multiple stakeholders
- Capabilities and confidence building that takes place using Adaptive Learning for Breakthrough results

Small-Scale Teams

At BR, small-scale teams typically consist of 5 to 10 people who are recruited from the area being targeted for improvement, such as equipment management or billing. To work effectively and achieve the needed improvement, the team must be cross-functional in nature. As Greenwood describes it:

> "If timely billing is a problem, then you must have accounting people, customer service people, and the people who do the billing in the field and who are directly involved with developing the solution to the problem. It is important that these people work not as departments, but as a problem-solving team working on a process".

In many cases, people at Burlington Railroad often started improvement activities without first being clear as to why the problem in question is being addressed. Often, people become uncomfortable, wondering whether or not the real or correct problem is being addressed, or whether there is not a more important issue to work on right around the corner. Problem areas become much clearer when one compares the actual circumstances to the business objective, the process requirements, or the customer expectations. The bottom line is that problem areas must be "discovered."

In our research, we found that one of the most interesting things about humans at work is the way we perceive things. On one hand, there is a strong need for stimulation and change. On the other hand, we resist change. A good example dates back to Russia at the turn of this century. The decision was made to change to the Gregorian calendar, which is in use throughout most of the civilized world today. The problem was that the peasants, thinking they were being robbed of several days and possibly years of life, rioted because of the adjustments to the dates. Hundreds were killed over this opportunity for adaptive learning gone awry.

In many ways, humans are like the animals who adapt to gradually changing conditions so well that threats to existence are frequently ignored. In industry, it's common for serious problems to be ignored for months and years, while trivial matters receive much attention. The serious problems have been with us for so long, we have adapted to them. On the other hand, if some change comes along to disrupt the status quo, we sit up and take notice, giving the squeaky wheel the most attention. The other reason why we don't notice problems when we should is that we are busy doing our "normal jobs." We just don't make time for problem recognition.

At BN, problem solving activities began with separating problems from symptoms, understanding the facts, organizing the data, and exposing them through data analysis. One of the most common mistakes in problem solving is trying to solve a symptom. Therefore, one of the key uses of the Macrologistics Quality Journey is to improve the process by insuring that we are working on the problem(s) and not the symptoms.

Urgent Measurable Goals

A goal setting process was used similar to Policy Deployment Goal setting, in which team goals were aligned to the business objectives.

In their world of work, they basically have three types of goals: individual, team/workunit and corporate (organizational). Their work goals serve as an essential part of conducting business activities successfully. Setting goals and reaching them provided the motivation and direction necessary for growth and success in important breakthrough areas of almost every team accomplishment, as summarized by Greenwood.

> "Urgent measurable goals are ones where the team works on things that matter now! The team knows exactly what to achieve and how success will be measured. Communicating without a desired outcome is like traveling without a specific destination. You may or may not end up in a place you really enjoy. Outcomes help us end up at the goal destination we want".

An outcome is the result we want, expressed in words and terms of the way we would like to see things happen, the way we want to feel, and what we expect to hear when we have the outcome. Goals and objectives are in a broader category than outcomes. Think of outcomes as goals that have been clarified and sharpened.

Goal setting for effective outcomes increases in importance within work teams. The Goal Setting Process encourages the team to identify and develop a sound understanding of customers' needs as well as business strategy. Teams who make Goal Setting a key part of their working processes encourage diversity of input, increased consensus, and firmer commitment between members.

Goal Setting for breakthrough outcomes has proven successful for the following reasons, which are common themes in the organizations we have researched over the years:

- Improves teamwork through a shared sense of approach and purpose.
- Heightens achieved performance by setting targets to be achieved.
- Identifies resource constraints or limitations.
- Distinguishes work load priority.
- Is challenging yet achievable.

Specific Project Example: The Moorhead Malting Facility

The objective was to implement an EDI based information system at the Moorhead Malting facility by 12/1/93. The breakthrough was achieved in an eight week period, producing dramatic results. The reason was that the team developed and implemented solutions that they were empowered to implement and the barriers were removed. According to Greenwood:

> "A breakthrough team does not produce studies that get analyzed and reviewed and analyzed again, and someday may even get implemented. No, a breakthrough team develops and implements solutions and usually gets results in as little as a week and usually in eight weeks or less".

Business Issues Involving Multiple Stakeholders to Improve Capabilities and Confidence Building

The key point in this step, both at BN as well as in many of the organizations surveyed, is for the members to put themselves as a team in the shoes of the customer and develop simple surveys to help scope out areas to focus the team on. Some of the areas to cover are:

- Quality
- Speed/timeliness
- Cost and functionality
- Availability and flexibility
- Responsiveness
- Durability
- Reliability

The main goal of all of this is to become a high-value organization and becoming one depends upon the rate of improvement. Customer data and business statistics helps speed up learning which is the central purpose of information — to speed up learning which speeds up improvement. Without change there is no improvement. However, change requires new knowledge and new knowledge requires learning. Thus, rapid improvement using structured business issue-oriented problem-solving techniques is all about rapid learning. According to Bill Greenwood:

> We use breakthrough teams to deal with business issues that typically involve the service to our customers. For example, providing 45 additional covered hoppers per month so a specific customer can expand into new markets in the Pacific Northwest was a challenge

for one of our team projects. As a result, the team members came away from the project with increased business literacy and a more in-depth understanding of the customers needs and wants. Team members also learn skills which can be applied to other parts of the person's job and to the way in which the railroad operates and continuously improves itself".

The ARCO Project Example

ARCO is a major customer of Burlington Railroads, and is served by the Cherry Point Refinery north of Seattle, Washington. Burlington Railroad handles movements of petroleum coke to Pacific Northwest Aluminum smelters. In 1993, BN had 155 of its covered hopper cars assigned to supplement ARCO's fleet to handle this business. The 155 cars were needed in service because of rail yard congestion and poor utilization of these covered hoppers.

A breakthrough team was formed and chartered to improve the situation. The identification and documentation of the team charter is the critical first step in the problem-solving process. Ironically, it is also the least understood and most frequently omitted or short-circuited. The nucleus of the charter revolves around the establishment of a performance promise that the team can consistently deliver in order to ensure success. According to Bob Lynch of QualTeam, flawless delivery on the performance promise to our customers and stakeholders depends upon seamless execution internally. At BN, seamless execution is the result of strong links between internal customers and suppliers. The charter enables the team to consider its role in helping to accomplish the greater purpose of the organization by considering these relationships.

In related research, we have found that the charter can often be viewed as a chain of objectives, which begins with a statement of the mission of the team, followed by the supporting goals. The length of the mission statement should be between 25 to 50 words and describes the core purpose of being for the team. The next section covers the Team Organization and Reporting Structure, followed by the Team Member Responsibilities. This is followed by the Procedures for each of the team's supporting purposes, as well as those items that are out of scope. Rounding out the Charter are the team Goals to be accomplished. The following are the key items typically covered within the charter framework:

1. Team mission and objectives.
2. Products, services, and/or information provided.
3. Synopsis of team processes, customers, and their valid requirements.
4. Competitive benchmarks, if known.
5. Supplier Requirements, if available.

6. Problem Statement and related symptoms
7. Charter Boundaries, including items not included

The Burlington Railroad/ARCO Breakthrough team found that many of the cars were safety stock needed to ensure that ARCO wouldn't run out of car supply. Thus, the opportunity was there to improve service to ARCO and reduce the number of cars required at the same time. Greenwood reports:

> "This looked like an ideal situation to test the breakthrough concept. But we needed a new way of doing this. The reason I said that is that in the past each department (train operations, freight car management, and marketing) would each do their own thing. All these employees were working as hard as they could, but the traditional BN approach lacked the cross-functional and cross-organizational cooperation required to fix this situation with ARCO. In other words, no one person was responsible and accountable for the ARCO business.
>
> All that changed in 1993 with the launching of Burlington Railroads's Account Leader initiative as a one-point contact. Let's take a look at what happened. The newly appointed Account Leader, Clint Watkins, is now accountable for coordinating all aspects of ARCO's business with BN. This includes freight car as well as utilization. Clint assembled the six appropriate people to be on the ARCO Breakthrough team.
>
> Three of the key people were: Doug Verity, the trainmaster at Bellingham, Washington; Gary Dunn, in charge of managing the covered hopper fleet; and Stu Gordon, the terminal superintendent in Everett, Washington. Without the full participation of these people, the project would have failed for the approach had to be cross-functional".

The seven-person team met and reviewed all the pertinent information and then took the key next step in making the Breakthrough approach work: the creation of razor-sharp goals. These are goals that are precise, clear, and timely. After identifying an area of improvement, goal setting for razor-sharp outcomes allows a team or an individual to create an incremental plan to increase their problem-solving and business outcome effectiveness. Goal setting illuminates the paths of solutions and helps measure successful endeavors. Razor-sharp goal setting improves performance by providing clear, tight targets to aim for. When teams adapt and meet relevant, razor-sharp goals, areas of concern improve systematically. According to Bill Greenwood:

"The ARCO Breakthrough team's goal, created by charter in April 1993, was to reduce excess inventory in Pool Assignment #5104 by 50% by June 1993. Notice how the goal is precise, clear, and timely. And now the hard part began: implementing it! In the past, this is the kind of situation that a marketing person would have tossed across the wall to an operating person to handle and report back on. With Breakthrough, the team is accountable to jointly achieve the results.

The team worked together with the ARCO people and reviewed in depth the shipping patterns, frequencies, and schedules. ARCO's customers who were the recipients also had their shipping patterns, frequencies, and schedules reviewed, thus extending their work throughout the entire value chain. Finally, Burlington's handling practices in distributing covered hoppers and in working through terminals were examined as well".

Goal Setting in Action

A standard "Goal-Setting System* relies on the use of action verbs to signify what will result from the completion of the goal. Many conpanies use the following list to help identify action verbs.

Accept	Aid	Assist	Check	Connect
Accomplish	Align	Audit	Choose	Conserve
Accumulate	Allocate	Authorize	Classify	Consider
Achieve	Analyze	Balance	Collect	Construct
Acquire	Appoint	Bargain	Combine	Consult
Activate	Appraise	Brief	Communicate	Control
Adjust	Approve	Budget	Compile	Convert
Administer	Arrange	Build	Complete	Cooperate
Adopt	Ascertain	Buy	Comply	Coordinate
Advise	Assemble	Calculate	Compute	Counsel
Advocate	Assess	Catalog	Conduct	Create
Criticize	Explain	Justify	Process	Secure
Critique	Express	Keep	Procure	Select
Decrease	Fabricate	Label	Program	Sell
Defend	Facilitate	Lecture	Project	Send

*This goal-setting method is based upon Edward de Bono's ideas set forth in *PO: Beyond Yes and No*, Penguin Press, 1972, and *Serious Creativity*, Harper Collins, 1992.

Define	File	List	Proofread	Serve
Delegate	Fix	Mail	Protect	Show
Delimit	Forecast	Maintain	Provide	Sketch
Deliver	Formulate	Make Available	Pull	Solicit
Depict	Forward	Make Happen	Purchase	Solve
Describe	Furnish	Manage	Rate	Sort
Design	Further	Manipulate	Read	Specify
Distribute	Gather	Measure	Recall	Staff
Detach	Get	Mediate	Receive	State
Detect	Give	Meet	Recite	Stop
Determine	Guarantee	Motivate	Recommend	Study
Develop	Guide	Name	Record	Submit
Devise	Grow	Negotiate	Recount	Suggest
Direct	Identify	Notify	Recruit	Summarize
Do	Implement	Observe	Regulate	Supervise
Draft	Improve	Obtain	Reject	Supply
Draw	Inform	Optimize	Remove	Survey
Edit	Initiate	Order	Render	Take
Eliminate	Inquire	Organize	Repair	Teach
Encourage	Inspect	Orient	Replace	Test
Endorse	Install	Originate	Report	Train
Ensure	Instruct	Overhaul	Request	Transfer
Erect	Insure	Participate	Require	Transmit
Establish	Interpret	Perform	Requisition	Tutor
Estimate	Interview	Persuade	Research	Utilize
Evaluate	Inventory	Pick-up	Resolve	Verbalize
Examine	Investigate	Pilot	Restrict	Verify
Exchange	Issue	Plan	Return	Weight
Execute	Itemize	Prescribe	Review	Withdraw
Expedite	Join	Present	Schedule	Word
Expand	Judge	Prevent	Score	Write
Experiment				

All of this was a new way of doing things at Burlington Railroad which directly resulted from the introduction of Breakthrough Goal Setting and set the stage for the way the organization continues to do business today. The Breakthrough team has a detailed plan covering how to resolve their situation to ARCO's benefit, as well as to BN. The team got the buy-in to make reductions in the fleet, but to make it work, they needed one more thing: precision execution! Bill Greenwood illustrates this point with a story about a man changing a flat tire:

> "It takes me about fifteen minutes to change a flat tire. Under normal conditions, this may seem fine. But let's say that I'm a driver at the Indy 500. Here, it takes a pit crew less than *fifteen seconds* to change a tire. What's the difference? First of all, at Indy, the customer is the driver of the race car and a fifteen minute change-over would be totally unacceptable. A few seconds can mean the difference between winning and losing.
>
> Second, at Indy there is a well-trained pit crew that acts as a team with each person knowing exactly what to do in a very precise way. Third, the team's focus is on the absolutely essential movements to change the tire, for there are no wasted moves in the pits at Indy. Fourth, every member of the team knows exactly what the goal is: to get that tire changed as quickly and safely as possible.
>
> Fifth, every member of the team has been trained to do his/her own job exactly right. And lastly, the team has practiced and practiced and practiced so that when it's race time, they perform each movement as precisely as they should. And with any luck, their customer — the driver — and the entire team will win the race".

While Bill Greenwood's example may seem a little extreme, it clearly illustrates what a highly focused, high performance team can do if they have razor-sharp execution. The racing team may not have called this process Breakthrough Thinking, but that is exactly the kind of thinking they used to figure out how to change a tire in less than fifteen seconds.

So how did the ARCO Breakthrough team do? Excellently well! Their initial efforts reduced 65 cars, or what amounted to 42% of the available cars from the fleet. These cars have been deployed to other business opportunities with ARCO and with other customers needing covered hoppers. According to Greenwood:

> "Remember, the razor-sharp goal was 50% and the team achieved 42%, or almost 85%, which was considered breakthrough results. Also, this was the first time through the Breakthrough process for this group. Why is this a big deal? Because these are the results of only one breakthrough ream. By the end of 1993, we had more than 200 breakthrough teams, all making productivity and cycle time improvements on a customer-by-customer basis. When you add up the results of the 200+ teams, they are in the tens of millions of dollars".

To accomplish all this required quite a bit of training. In the first year alone, more than 400 Burlington Railroad people had formal training in the Breakthrough process and how to use razor-sharp goal setting effectively.

Lessons Learned

At Burlington RR, they learned that you must include on the team from the start all the people with a major stake in the project's success. This means that they must be part of setting the razor-sharp goals and developing and implementing the plan. Second, make sure that there is a razor-sharp goal. Without it, the team will flounder and not stay focused. Third, make sure the team has a master workplan. The masterplan with a clear understanding of the rollers each team member plays is essential to success. Finally, empower the team to do what needs to be done, get them the resources needed, and then get out of the way. Bill Greenwood summarizes in this manner:

> "The need for breakthrough thinking exists because nothing in this world stands still. And it never will. Our customer requirements, our markets, and our technology keep evolving. And that's good, for it creates improvements to our products and services. This constant change makes us focus on the essential parts of providing our services; and on which parts add value and which ones don't; and on new ways to do things — breakthrough ways!
>
> The introduction of the Breakthrough Learning process has been very successful and rewarding experience. I also think that we have just scratched the surface of the potential that breakthrough thinking has in Macrologistics."

Rubbermaid's Simple Strategy: Innovate 6.2

In 1993, after being number two in *Fortune* magazine's most-admired-company contest for five of the previous six years, Rubbermaid (finally,) became number one. The company is a veritable Juggernaut when it comes to putting out new products: 365 a year — that is, one a day. Headed by Stanley Gault from 1980 to 1991, the company enjoyed unprecedented growth in profits (an average increase of 14% a year) and stock appreciation (an average of 25% a year). When Gault retired in 1991, only to move on shortly thereafter to become CEO of Goodyear (he's still chairman of the board at Rubbermaid), he was replaced by Walter Williams, who resigned after 18 months.

Now the firm is run by Wolfgang Schmitt who has established high objectives for the firm but will continue to pursue its long-established innovation strategy to achieve those objectives. Schmitt wants Rubbermaid to enter a new-product category every 12 to 18 months (most recently it has introduced hardware cabinets and garden sheds); to obtain 33% of its sales from products introduced within the past five years; and to obtain 25% of its revenues from markets outside the United States by the year 2000, an increase of 7% over the current 18%.

Rubbermaid excels in making mundane items seem interesting and functional, and it also makes them profitably. Each year it improves over 5000 existing products or creates totally new ones. Its product line includes mailboxes, window boxes, storage boxes, toys, mops, dust mitts, snap-together furniture, ice cube trays, stadium seats, spatulas, step stools, wall coverings, sporting goods, dinnerware, dish drainers, laundry hampers, and many other utilitarian products. However mundane those items may be, Rubbermaid's engineers hover over their products as intently as would General Dynamics' engineers over an F-111 fighter. It is this serious approach to what others dismiss as trivial that has helped make the firm so successful.

Most of Rubbermaid's new products come from twenty cross-functional teams, each with five to seven members (one each from marketing, manufacturing, R&D, finance, and other departments, as needed). Each team focuses on a specific product line so that someone is always thinking about key product segments. But innovation doesn't stop with the teams. Individual employees are geared toward creating new products as well.

Rubbermaid has taught its employees to think in terms of letting new products flow from the firm's core competencies — the critically important things it does well. It encourages its managers to find out what's happening in the rest of the company, continually looking at processes and technologies. For example, while running a different Rubbermaid subsidiary, Bud Hellman toured a Rubbermaid plant that made picnic coolers. As he watched the plastic blow-molding equipment, he realized that he could use that process to make a line of durable, lightweight, inexpensive office furniture. Within a couple of years that line accounted for 60% of the furniture division's sales.

Top management often contributes ideas as well. When CEO Schmitt and Richard Gates, head of product development, toured the British Museum in London in 1993, they became extremely interested in an exhibit of Egyptian antiquities. They came away with eleven specific product ideas. Gates says admiringly of the Egyptians, "They used a lot of kitchen utensils, some of which were very nice. Nice designs."

Typical of Rubbermaid's attention to detail is its approach to customer relations. In the Customer Center, which hosts 110 major retail customers a year (including the biggest, Wal-Mart, which accounts for 14% of the firm's sales) the pitch is always the same: "Let us help you sell more; we've got what consumers want." At the end of each presentation, the customer sees many new products. Then it's on to the War Room, where the deficiencies in competitors' products are demonstrated. Then it's on to the Best Practices Room, where retailers see the best in product mixes and displays. The idea is to establish a store within a store — that is, to put a Rubbermaid store in the retailer's stores.

Rubbermaid does no market testing, although it does hold focus groups. Schmitt doesn't believe in testing. "We don't want to be copied. It's not that much riskier to just roll it out. Plus, it puts pressure on us to do it right the first time," he states. Flops do happen occasionally, but the firm has a remarkable 90% success rate with new products and tolerates the few failures that occur in the name of taking risks.

© James M. Higgins. *Sources:* Alan Farnham, "America's Most Admired Company," *Fortune* (February 7, 1994), pp. 50–54; Brian Dumaine, 'Closing the Innovation Gap,'" *Fortune* (December 2, 1991), p. 57.

Hitachi Innovates Like Nobody's Business 6.3

With $62 billion in sales, Hitachi Limited produces nearly 2% of Japan's yearly gross national product. It has 28 factories, 800 subsidiaries, and 320,000 employees. It annually sells $9 billion worth of consumer electronics products, and yet it also sells $20 billion worth of power plants, generators and robots. It also has a computer chip division that annually sells more than Motorola, Intel, and Sun Microsystems combined. It is singly responsible for 6% of Japan's total R&D expenditures. It is Japan's largest patent holder and has been at the top of the U.S. patent list for most of the past ten years. Of its rivals in size, General Electric, Matsushita and IBM, only Hitachi has interests in more than three of these business areas: computers, chips, software, consumer electronics, power plants, transportation, medical equipment, and telecommunications. It has interests in all eight areas. Thus it is capable of undertaking huge projects, such as a national maglev logistics transportation system, that literally no one else could undertake.

Hitachi's success story begins with product R&D, but goes beyond new products to include processes in all areas of the firm helping drive costs down, down, down. Facing increased competition and sagging demand in many of its business areas such as mainframe computers, Hitachi has turned to spartan costcutting to maintain its market positions. It is advancing its vision of the future through a sophisticated, innovative information network where technology fusion is a frequent result. Sharing information among sister units is seen as paramount to new product development and cost cutting.

Hitachi's 63-year-old president, Tsutomu Kanai, is just the non-conformist to lead this firm. He has a vision for the firm, one that takes advantage of its diversity in an integrated way. Thus the firm is pursuing integrated complex projects with zest. Because it has huge cash reserves and has invested in basic businesses that survive recessions well, Hitachi has been able to protect its product R&D stream more than most firms. "Basic science is something we will never sacrifice," observes Kanai.

Hitachi, unlike other Japanese firms, has a loose, decentralized management structure. Renegades are encouraged by this system. One such renegade is Yasutsugu Takeda, Hitachi's top R&D administrator. He was undaunted by the fact that none of Hitachi's factory managers were interested in making several products his labs had developed. So he created catalogs of these products, lined up customers, and then went to Hitachi factory managers and convinced them that they should make these products after all. Decentralization is the name of the game at Hitachi. The 28 factories are run more as separate businesses than as factories. It's often hard to tell exactly who runs the firm. There are ten managing directors, and clearly, Kanai is President. But the firm moves as a large group of equals might move, more than a pryamidal organization run by a powerful CEO would react.

Hitachi is working on new products at various distances into the future. In 3 to 5 years, they expect to have neural networks and multimedia office products. In 5 to 10 years, they foresee hand-held computers that accept voice commands and exchange data over radio waves. In 10 to 20 years they expect to produce computers with 100 times the power and 10 times the speed of today's models. Over 20 years from now, they expect biocomputers that can organically repair themselves. Innovation is truly the focus at Hitachi.

 Source: Takeo Imori, "Hitachi: Too Little, Too Late?" *Tokyo Business Today* (December 1992), pp. 12–13; Neil Gross, "Inside Hitachi," *Business Week* (September 28, 1992), pp. 92–100.

Eastman Chemical Reinvents Its Innovation Process 6.4

Although Eastman Chemical Company (ECC) had long recognized the importance of product and process innovation, it was not until 1988 that the firm became acutely aware that its innovation process was not sufficiently productive. A major study was launched to determine what improvements were needed; the firm then began developing strategies for implementing those improvements.

Before 1988, ECC's innovation process began with identification of a customer's needs. This effort included not only external customers but also internal customers such as manufacturing. This step was followed by several others: discovery-driven research, R&D efforts to develop potential solutions, a desirability/feasibility/capability study, designation of the project as an innovation concept, an applied research project, and finally, commercialization of a new product.

From the beginning of the process to the end, innovation was seen as an R&D function, not as something involving the whole company. More specifically, this meant that the resource allocation process was especially burdensome and that the new-product development process managed effectively. There was little participation by marketing and manufacturing and little reward for such participation when it did occur. Technology transfer from the lab to the market was often weak. There was too little interaction between R&D and the firm's business centers in determining market needs. Project managers were not well trained in project management or innovation management.

ECC set out to change these conditions. It established seven goals. The first aim was to make everyone at ECC an active participant in the innovation process. The second was to increase the participation of business centers. Third, ECC wanted to make certain that all functions gave their commitment to innovation from the beginning of a project. The fourth goal was to move projects through the process more quickly and with a higher success rate than in the past. Fifth, the need for a particular project was to revalidated throughout the project's life. Sixth, ECC would identify new business opportunities and create projects to take advantage of them. Finally, a system for managing and continually improving the innovation process would be created.

To achieve these objectives, ECC divided the innovation process into four parts:

1. **Needs identification** included systems for identifying, validating, and communicating product and process needs.
   - Multifunctional teams would be used in separate product/market categories to carry out these steps.
   - Business centers would review and support high-impact and other significant validated needs.
   - A futures technology and market research function would be established.
2. **Concept development** involved generating ideas and developing viable concepts to be implemented.
   - Informal multifunctional teams devoted to concept development early in the innovation process would be formed.
3. **Innovation concepts/applied research** comprised ECC's system for new product and process concepts being commercialized via a structured project management system.
   - An improved project management system would be implemented.
   - A company-wide system for assigning priorities to innovation projects would be implemented.
   - A full-time project manager would be appointed for high-priority projects.
4. **Market development** included market evaluation, introduction, and sales development for the new product.
   - For new-product projects, a market development team consisting of representatives from appropriate marketing organizations would be formed.
   - Appropriate measures of innovation would be established for each functional and business area.
   - Certain administrative processes would be streamlined.
   - A process facilitator would be appointed to oversee the entire innovation process.

After three years the program had scored several significant measurable improvements, notably in project completion cycles, the value of products produced, and the success rate of new products. From a qualitative standpoint, much of the change was in the way the process was managed — for example, in the involvement of all major functions and in the systems for assigning priorities and allocating resources to projects.

 Source: Jerry D. Holmes, Gregory O. Nelson, and David C. Stump, "Improving the Innovation Process at Eastman Chemical," *Research-Technology Management* (May/June 1993), 27–35.

Samsung Group Seeks More Innovation 6.5

Lee Kun-Hee, chairman of Korea's Samsung Group, is proud of his firm's sales of $54 billion a year, but he also wants Samsung to become a nimble global competitor. Believing that it has to achieve greater innovation and quality, he is grafting many Western management ideas onto the firm's Confucian-based hierarchy. His actions, dubbed the Second Foundation, constitute a remarkable change in organizational design, culture, and leadership style. Lee wants Samsung to become one of the world's ten largest technological powerhouses as well as the world's fifth-largest electronics firm, by the turn of the century. Only if the Second Foundation succeeds will the firm achieve those goals. Some observers feel that all of South Korea's competitiveness may be at stake, as other Korean firms will either emulate Samsung's successes or avoid its failures.

Samsung Group already boasts Korea's most industrious and able management teams, but the Confucian organizational culture perpetuates an authoritarian management style that often stifles creativity and innovation. Lee wants to change that style by delegating more authority, encouraging more risk taking, rewarding innovation, transforming the organization. In effect, he wants to turn the hierarchy upside down, abandoning the inward-looking approach to management established by his late father, Lee Byung-Chull. "I am telling them to change everything except how they treat their families," says the 51-year-old chairman.

Recognizing that change begins at the top and that he himself had to set the example, Lee slowly began delegating authority to his subordinates. To force managers to make decisions, he often worked at his guest home, refusing to take their calls or accept their visits. Still they resisted the changes. In 1993 Lee took his entire top management group to Los Angeles to show them how customers ignored Samsung's products. Still there was little change. Later that year on a flight to Frankfurt, Germany, a trusted Japanese consultant reported that the Samsung Design Center was being poorly run and no one at the lab seemed to care about changing it. The normally mild-mannered Lee hit the ceiling. He called Seoul and in a taped conversation, yelled at his senior executives for an hour. He then ordered copies of the tape distributed to all top managers. Shortly thereafter, he issued what came to be known as his Frankfurt Declaration: "Quality first, no matter what." He summoned hundreds of managers in groups of twenty to forty to Frankfurt for round-the-clock meetings. Eventually, all of the company's top managers went through sessions with him. Lee used the opportunity to demonstrate the poor market response to the company's products, but he spent most of the time discussing competitiveness, marketing, quality, and training. Lee also imposed numerous changes on the firm, such as dramatically changing work hours.

His message is that everyone in the firm must learn things that may be alien to its culture: otherwise Samsung is doomed to failure. He expects results, but he knows they'll be slow in coming. He is determined to make his firm more innovative and quality oriented in order to make it more competitive, but he recognizes that not all of his top managers will be able to handle change well. He expects that 5% will leave the company; 25 to 30% will have reduced responsibilities; and only 5 to 10% will receive increased responsibility.

 Sources: Lakmi Nakarmi and Robert Neff, "Samsung's Radical Shake-up," *Business Week* (February 28, 1994), pp. 74–76; Ed Paisley, "Management: Innovate, not Imitate," *Far Eastern Economic Review* (May 13, 1993), pp. 64–68.

Stranco: Incremental vs. Radical Innovation 6.6

Stranco Incorporated was founded in 1970 to manufacture products for water chlorination. Its products incorporate new technologies created for this market by founder Frank Strand. To speed new-product development and growth, Strand encourages two types of innovative thinking: original thinking or radical innovation, and systematic creative thinking or incremental innovation. Major breakthroughs come from the original thinkers, while ongoing improvements come from the creative thinkers. The following table compares the two types of thinking.

Original Versus Creative Thinkers

Original	*Creative*
■ More right-brained	■ More left-brained
■ Approach problems from new angles	■ Systematic approach to problems
■ Tend to be loners	■ More sociable, competitive
■ Operate in messy environments	■ Love results, progress, feedback
■ Eccentric, extreme mood swings	■ Neater, more methodical
■ Well-developed sense of humor	■ More stable personality

Stranco has discovered that each type of thinking requires a distinct management style to make it most effective. Stranco's radical thinkers need a free rein. They don't like budgets or deadlines; they work best in a trusting, respecting environment. Conversely, the incremental thinkers respond to assignments and follow-ups. These employees can be managed through weekly technical meetings, problem definition, and milestone reviews. The following table compares the characteristics of the management style that Stranco has found to be best for each type of thinking.

Managing Radical and Incremental Thinkers

Radical	*Incremental*
■ Stimulate through challenges and puzzles	■ Set systematic goals and deadlines
■ Remove budgetary and deadline constraints when possible	■ Stimulate through competitive pressures
■ Encourage technical education and exposure to customers	■ Encourage technical education and exposure to customers
■ Allow technical sharing and brainstorming sessions	■ Hold weekly meetings that include key management and marketing staff
■ Give personal attention — develop relationships of trust	■ Delegate more responsibility
■ Encourage praise from outside parties	■ Set clear financial rewards for meeting goals and deadlines
■ Have flexible funds for opportunities that arise	
■ Reward with freedom and capital for new projects and interests	

© James M. Higgins. *Source:* Harry S. Dent, Jr., "Growth Through New Product Development," *Small Business Reports* (November 1990), pp. 30–40.

IKEA: Marketing Innovation Drives Total Business Redefinition 6.7

IKEA was transformed from a small Swedish mail-order furniture operation into the world's largest retailer of home furnishings after top management totally redefined the furniture retail business. In an industry where few companies move outside their own countries, IKEA has created a global network of over 100 stores. In 1992, 96 million people visited these stores and purchased $4.3 billion in goods. The key elements of IKEA's successful strategy are well known:

simple, high-quality, Scandinavian design; global sourcing of components; knock-down furniture kit that customers transport and assemble themselves; huge suburban stores with plenty of parking and amenities like coffee shops, restaurants, even day-care facilities. Low prices, ranging from 25 to 50% below competitors are also an essential ingredient, created by a combination of low-cost components, efficient warehousing, and customer self-service.

But if one looks beyond the obvious strategy one finds the real innovation. IKEA attains low costs and translates them into low prices because it fundamentally redefined the business. First, it redefined the relationship with the customer. The new relationship promises the customer low prices and high quality if the customer will perform two tasks that are normally performed by the manufacturer and the retailer: assembly and delivery. Because IKEA wanted to make it easy for the customer to take on this new role, every aspect of the company was defined accordingly.

For example, each year IKEA prints 45 million catalogs in ten different languages. Each catalog lists between 30 and 40% of the firm's 10,000 products. Most important, each provides a script describing the role of each participant in the company's business system. When customers enter a store they are given a catalog, tape measures, pens, and notepaper to help them make choices without the aid of a salesperson. Furniture items come complete with simple, readable labels with descriptions of the dimensions, colors, and materials in which they are available, instructions for care, and the location in the store where they may be picked up. After payment, customers cart off the unassembled items. Car roof racks are available. In these ways IKEA has redefined the way value is created. "IKEA wants its customers to understand that their role is not to *consume* value but to *create* it."

In similar ways, IKEA has redefined its relationships with suppliers. For example, it seeks long-term commitment from highly qualified, globally based suppliers, which must meet rigorous tests before becoming members of the IKEA team. And it has created a special unit, IKEA Engineering, to provide technical assistance to suppliers. Finally, IKEA also reinvented its internal processes, which mirror its relationship with its customers and suppliers. For example, its logistics process deals with global sources to assemble widely dispersed components. The back of a chair may come from Poland, the legs from France, and the screws that hold it together from Spain. For this to be a financially sound strategy high-volume sales are required. At the center of this strategy are fourteen warehouses linked to retail stores in computer-integrated networks. These warehouses are in fact system integrators, relating supply to demand, holding down warehouse inventories, and keeping stores supplied only as needed.

IKEA's success has occurred because it invented a new way for value to be created rather than trying to systematically create value along a given value chain.

© James M. Higgins. *Source:* Richard Normann and Rafael Ramlerez, "From Value Chain to Value Constellation: Designing Interactive Strategy," *Harvard Business Review* (July-August 1993), pp. 65–77.

Shiseido Meditates for Creativity 6.8

At the heart and soul of Japanese organizations is their consensus approach to management. It works well when they are pursuing continuous improvement in processes and products, but not so well when it comes to creating new products, markets, and whole new businesses — activities that will be vital to corporate success in the 21st century. Thus, Japanese organizations are attempting to make their cultures more receptive to bold new ideas. A few, such as Hitachi and Sony have always been receptive to new ideas; but most have not.

Japanese firms are trying different approaches to develop creativity. Omron Corporation, a manufacturer of electronic controls, holds a monthly *juku,* or cram school. In a process is designed to stimulate creativity, middle-level managers take on various roles, such as a 19th-century warlord, Formula One race-car driver, or private detective. They then tackle problems, thinking and planning as one would in those roles.

At Fuji Film, senior managers are asked to study offbeat subjects such as the history of Venice or the sociology of apes. Useful insights into business problems may occur as a consequence of the new perspectives gained in this manner. Similarly, employees at Shimizu Corporation, Japan's largest construction firm, spend several days each year at company retreats, playing games that force them to tackle "impossible" problems such as getting back from the moon in a damaged spacecraft.

Shiseido, Japan's largest cosmetics maker, recently undertook a series of four-day seminars at resorts on the snowy shoulders of Mount Fuji. It was an ambitious attempt to use philosophical introspection to change corporate culture. The seminars were undertaken despite the company's continued financial success during a period of recession. The firm's president, Yoshiharu Fukuhara, expressed his concern this way: "Our company cannot be like a military troop where the president gives an order and then everyone rushes to do it. Companies like that will not survive in the next era." His aim is to create a company in which individuals set their own goals and find ways to achieve them. This is a revolutionary approach for a Japanese firm.

The four seminars were titled "Time and Space," "Expression and Language," "Body and Soul," and "Beauty and Truth." They consisted of a mixture of lecture, discussion, self-disclosure, introspection, and exercises. For example, "Body and Soul" began with a lecture on the words "human being." It was presented by Seigo Matsuoka, an independent consultant who viewed his role as one of bridging the worlds of technology, business, and culture. Next there was an open discussion of progress toward solving personal problems that had been revealed at the previous seminar. This process was not easy for men who had spent their entire lives stifling their personal feelings for the good of the company. The next day, discussions of personal vulnerability continued, followed by a lecture on the organization as a living system, delivered by Hiroski Shimizu, a professor of pharmacy at Tokyo University. In the evening the group studied goldfish in a bowl and drew parallels to the firm in its environment. On day three, President Fukuhara addressed the group. After showing a video of his favorite Kabuki actor and a film of an economist talking about the intricacies of corporate culture, he disclosed some of his own vulnerabilities and presented his philosophy for the company.

After lunch the participants listed the physical quirks of other members of the group. These were then discussed in a session led by Seiko Ito, a psychologist. Then body language became the focus of the discussion. Three men were asked to act like Americans. They did so by moving about in an exaggerated fashion, flailing their arms. Another leader provided exercises related to body and motion. For example, participants were asked to balance an egg on end (About half succeeded in doing so.)

The fourth day began just before dawn with a meditation led by Zen master Ryomin Akizuki. Reflective discussion followed. After lunch the group was asked by consultant Matsuoka to summarize what they had learned. Most acknowledged that they had achieved a different perspective, but it was hard for them to list any actions they would take to make specific changes in their lives.

This firm's experience is typical of that of many firms, in the United States and Europe as well as the Pacific Rim, in introducing more creativity into their cultures. The process often involves trial and error when you approach it from a philosophical perspective. Shiseido's program may not work elsewhere — different approaches have to be used in different cultures. As a human resources manager at Fuji Film puts it, "You can't just tell your employees, 'Be creative.'" Firms have to create an environment that is conducive to greater individualism, something that most haven't done in the past.

 Source: Emily Thornton, "Japan's Struggle to Be Creative," *Fortune* (April 19, 1993), pp. 129–134.

IBM Changes its Culture in Order to Reinvent Itself 6.9

When Lou Gerstner became CEO of IBM, he quickly shelved the three "basic beliefs" that had guided the firm for decades: pursue excellence; provide the best customer service; and above all, show respect for the individual. In their stead, he proposed the following guidelines:.

1. The marketplace is the driving force behind everything we do.
2. At our core, we are a technology company with an overriding commitment to quality.
3. Our primary measures of success are customer satisfaction and shareholder value.
4. We operate as an entrepreneurial organization with a minimum of bureaucracy and a never-ending focus on productivity.
5. We never lose sight of our strategic vision.
6. We think and act with a sense of urgency.
7. Outstanding, dedicated people make it all happen, particularly when they work together as a team.
8. We are sensitive to the needs of all employees and to the communities in which we operate.

Longtime IBM employees and managers were shocked and dismayed. Thomas J. Watson, Jr., the son of IBM's founder, had been preaching the three basic beliefs for decades. They were ingrained, practically a mantra to be recited by every member of the company. Watson had said that everything at IBM could be changed except these basic beliefs. Most galling was the relegation of respect for employees to the last place on the list.

Gerstner also expected IBM employees to act like owners. Lee Conrad, a dissident at an IBM plant in Endicott, New York, retorted in a newsletter to colleagues, "Nice words, Lou, but it's hard to think as an owner when we're treated like hired hands." Gerstner thus experienced at first hand the difficulty of changing an organizational culture that has been firmly entrenched for seventy years.

The first thing Gerstner did when he came to IBM, was to attack the firm's financial weaknesses. In July 1993 he took a $8.9 billion pretax charge and announced cuts of 35,000 jobs. In March 1994, he announced his new strategy. Next he has to tackle the heart and soul of the firm, its culture. Impatient with IBM's culture of arrogance and lethargy, he has informed employees that they should not feel entitled to their jobs. He balks at the "uniform" of blue suit and white shirt and dislikes the overhead transparencies that have traditionally been used for presentations. He holds town meetings during which he answers employee's questions. He has initiated new personnel policies and a new performance appraisal system and has completely changed the IBM training program, giving it a clearer focus on customers and quality. He is revamping the sales force to make its members business problem solvers, not just salespeople. Through various symbolic gestures and public statements, he has let it be known that research and development are still essential to IBM. But perhaps most important, through firings and resignations of those who opposed him, he has made it clear that he won't accept anything less than full compliance with his goals.

Sources: Ira Sager, "The Few, the True, the Blue," *Business Week* (May 30, 1994), pp. 124–126; Peter Coy, "Is Big Blue Still Big on Research? You Bet." *Business Week* (May 16, 1994), pp. 89–90; Laurie Hays, "Blue Period: Gerstner Is Struggling as He Tries to Change Ingrained IBM Culture," *Wall Street Journal* (May 13, 1994), pp. A1, A8.

Reinventing The Chrysler Corporation 6.10

In 1989 Chrysler Corporation was faced with a series of problems, including an aging line of cars, a bulging cost structure, and declining profits. Top management, spearheaded by Lee Iacocca, recognized the need for a completely new strategy if the company was to be competitive. Chrysler had to develop new platforms (basic car bodies around which models are built), achieve superior advanced designs, reach the highest levels of quality to entice the typical U.S. buyer of Japanese cars, and cut unnecessary costs — and it had to do all this as quickly as possible. The phrase "reinventing the Chrysler Corporation" became synonymous with the effort to achieve a competitive advantage. It quickly became evident, however, that to make the new strategies effective the firm also had to drastically change its structure, leadership style, management systems, and organizational culture.

Chimneys and Platform Teams

Among the first problems tackled was that of getting new products to market faster with higher quality and higher levels of customer satisfaction. At Chrysler, as at most U.S. corporations, design had historically been isolated from other functional units. The product went from design to engineering to procurement and supply to manufacturing, and finally, to marketing and sales. Each of these functional units had its own bureaucracy, which did not communicate with the others unless they balked at proposed plans; then top management had to settle the matter. This structure led to constant bickering. Each functional unit was like a chimney, billowing its own smoke upward to top management. Designs that were created and manufactured in isolation were forced on customers by the sales force, whose job was to sell the product whether customers liked it or not. Meanwhile the finance department was trying to forecast costs and revenues in isolation from the normal flow of information.

Chrysler decided to adopt cross-functional development teams, called *platform teams*. Four teams were created: large car, small car, minivan, and jeep/truck. Representatives from each of the functional departments, plus customers, were integrated into the product development, manufacturing, and marketing process. Finance was included in the information loop, but the teams were charged with bringing products to market within precise cost levels. The teams were given a high degree of autonomy to achieve their objectives, eliminating the need for management to continually settle disputes.

To facilitate product development, the work of the teams, and the cross-functional nature of their operations, Chrysler officially opened its $1 billion Chrysler Technology Center (CTC) in the fall of 1991. Each platform team has its own floor in the CTC. Team members use shared databases and communication systems. Suppliers are involved throughout the design process. Through focus groups, questionnaires, and interviews, customers are also involved. The CTC includes a manufacturing facility where prototype manufacturing processes can be devised at the same time that a new car is being developed in order to speed manufacturing and improve quality. The platform team works in conjunction with assembly line workers to create the most efficient manufacturing process possible.

Changes in Macro Structure, Leadership Style, Management Systems, and Culture

One of the major changes in structure that occurred during this period was the empowerment of employees to make decisions that were relevant to their jobs. In the past, the supervisor typically made all the decisions. But management recognized that this approach was not entirely effective. In the new structure the supervisor would become a coach and facilitator. Part of the effort to cut up to $4 billion in costs depended on employees making suggestions that would eliminate unnecessary tasks.

The changes in Chrysler's structure, style, and culture are reflected in the company's New Castle, Indiana plant. Once targeted for closing, the plant went from losing $5 million a year in 1988 to saving the firm a net of $1.5 million in 1991. This impressive turnaround was accomplished by empowering employees, making sure they knew that change was necessary, and creating a learning environment. Information systems were updated to provide employees with the information they needed in order to make the right decisions.

Results

The results of those changes have been impressive. Chrysler posted significant profits in 1992 and 1993. The price of its stock soared from a low of $10.50 per share in late 1991 to $61.00 per share in February 1994. The profits and stock price reflect Chrysler's tremendous success with the Dodge Viper, an exciting and pricey new sports car; the latest Jeep Cherokee; and the LH series of cars, which entered the medium-price market against European and Japanese competitors and proved extremely successful. Trendy, with many popular features, the cars are so successful that Chrysler's only problem is how to manufacture them fast enough. Additional new models anticipated for 1994 and 1995, including the Dodge Ram truck and the small Neon, are also receiving positive reviews.

 Source The Chrysler Corporation, "Reinventing Chrysler," a videotape, (Detroit: The Chrysler Corporation) 1993; and recent stock market reports.

Norfolk Southern: On The Right Track 6.11

The railroad industry was mired in a depressed condition just fifteen years ago. But with productivity increases that averaged 157% over that time frame, increased use of technology, significant capital investment, quality management programs, and a stronger customer focus, the railroads have made a comeback. At the head of the herd is Norfolk Southern, whose lone black stallion advertising symbol epitomizes its leadership role in the industry. Norfolk Southern, the nation's fourth largest railroad, was one of only two of the nation's major railroads to make enough money to cover its cost of capital in 1992. In 1993, it made $772 million on sales of $4.5 billion.

Chairman David Goode is steering the company toward a strategy of marrying the traditional efficiencies of railroad freight hauling with the needs of today's just-in-time economy, requiring rapid, flexible, and dependable service. To make that strategy work requires a significant amount of innovation. Fortunately, Norfolk Southern is a firm with a recent history of innovation, and a firm still concerned with innovation. Innovation in managing boxcar loading, sorting cars in various hubs, and eliminating the need for certain hubs on some routes, for example, has enabled Norfolk Southern to garner a significant share of both the automobile delivery business and the truck trailer hauling business. The firm can, for example, guarantee that a delivery will be made at a precise minute to the Atlanta Ford Motor Company assembly plant. This enables Ford to use boxcars as inventory bins, cutting their own inventories and making Just-In-Time inventorying possible. There is some room for leeway, but being late more than a few minutes would idle some 2400 workers at the plant.

Significant technology investment has gone into communications technology enabling better tracking of the various types of railroad cars, their loads, and where they are in the system relative to where they need to be. Another focus of information technology has been to improve customer satisfaction with railroad operations, for example, in bringing new service ideas to customers. Innovative quality and safety management programs have helped improve efficiency while cutting accidents by two-thirds.

Norfolk Southern's focus on innovation springs from a series of three- and four-day seminars on creative problem solving attended by virtually all upper and middle managers and many professional staff members from 1981 to 1988. The basic creative problem solving model was used as the focal point for the course. Traditional creativity processes such as brainstorming were taught, along with less traditional, but very useful processes such as excursion. In 1982, Southern Railway merged with Norfolk and Western Railway. The creative problem-solving course provided an extra benefit to the new firm by offering a way for the two merged firms' managers to learn to solve problems together. That focus on innovation and collaboration has remained with the firm.

In 1990, the firm began to think in terms of developing an innovation research program. "Throroughbred Innovation" was formalized in early 1993 with the creation of the Innovation Research Group. The Group's specific mission is to locate new technology that can be applied to railroad operations. Therefore, most of its efforts are aimed at process innovation. The 1993 objective for the Group was very modest — obtain $1 million in savings. But within five years, the firm expects to save a minimum of $40 million a year from the Group's efforts. Some of the innovations produced include: a new type of trainman's lantern, a strong yet lightweight coupler knuckle (which couples cars together), Timbrex crossties made of recycled materials, and high-hardness wheels. The firm is also examining the possibilities of fuel injection for locomotives and alternative fuels such as liquefied natural gas.

© James M. Higgins. *Sources:* David Hage, "On the Right Track," *U.S. News & World Report* (March 21, 1994), pp. 46–53; William G. Vantuono, "Productivity Propels Rolling Stock Buying Decisions," *Railway Age* (January 1994), p. 15; William G. Vantuono, "C&S: State-of-the-Art Improves State-of-the-Art Railroad," *Railway Age* (January 1994), p. 40; Michael A. Verespej, "Better Safety Through Empowerment," *Industry Week* (November 15, 1993), pp. 56–68; Gus Welty, "NS's 'Thoroughbred' Innovators," *Railway Age* (August 1993), pp. 95–96; Robert J. Bowman.

7 Just-In-Time (JIT)/JIT II™

Inventory in the value chain is a major contributor to poor quality and longer cycle time! For a breakthrough related to cycle time to occur, there needs to be an alignment using the concept of JIT, without which cycle time reduction cannot be dramatically impacted. It also requires the total committment between suppliers and logistics service providers using the concept of JIT II™.* This is a major principle of Macrologistics Management. Because inventory is any idle goods that is held for future use, it often hides many quality problems. In the case of a bad part, there is usually a backup on hand. With the aging of the inventory and larger lot sizes, proper feedback is masked by the days, weeks and months that have gone by. Thus, the harder it is to correct quality problems because the data trail is stale. In addition, inventory causes excessive material handling, which contributes to cost and does not add value to the product.

> The overall goal is to reduce inventories to as close to zero as possible by producing only enough work units to keep the next work station in a production process in operation. "The chief logistics managers are concerned with using time more effectively in the logistics channel."
>
> Bernard J. LaLonde, Professor of Transportation,
> Ohio State College of Business

---

* JIT II™ is a registered trademark of the BOSE Corporation; developed by Lance Dixon.

.Just-In-Time (JIT) is a Japanese management technique to inventory control and materials management that has as its aim the complete elimination of waste, including "unnecessary inventory" and scrap in production.[1] Schronberger describes JIT as "a quality and scrap control tool, as a streamlined plant configuration that raises process yields, as a production line balancing approach, and as an employee involvement and motivational mechanism."[2]

JIT Overview

JIT production and inventory control was invented and perfected by Toyota Motors under the direction of Yaiichi Ohno, who originally began on a trial basis in 1952 with the simple elimination of waste. Ohno first gained the insight for JIT from a visit to an American supermarket after WW II, at a time when Japanese manufacturer's access to raw materials and foreign exchange was severely limited. During his visit, Ohno observed how the U.S. supermarket experienced a constant flow of goods, avoiding both excess and out-of-stock conditions. He proposed this approach to Japanese manufacturing through the concept of "*kanban*", which literally means "using a signboard," or a communication tool-based card that serves as a quantity indicator for incoming inventory. When the inventory is depleted, the "kanban" card is returned to its point of origin for the replenishment of the order cycle.

The following is an extract from "Kaizen", by Masaaki Imai, describing the Toyota Just-In-Time production system, which the author likens to an example of management-oriented Kaizen:[3]

> "Toyota's Motomachi plant has a long line of trucks waiting outside the plant with full loads of automotive parts and components for the assembly line. As soon as one truck comes out at one end of the plant, another goes inside. There is no warehouse for these parts. Upholstered seats, for example, are fed to the production line directly from the back of the truck! The Toyota production system is now attracting great attention in Japan and abroad, for Toyota is one of the few companies to have survived the oil crises and still maintained a high level of profitability. There is much evidence supporting Toyota's success...which stems from a production system based upon two main structural features: the Just-In-Time concept and "*jidohka*" (automation).[4]

The concept of JIT means that the exact number of required units is brought to each successive stage of production at the appropriate time. Putting this into practice meant a reversal of the normal thinking process. Ordinarily, units are transferred to the next production stage as soon as they are ready. Ohno, however, reversed this, so that each stage was required to go back to the previous stage to pick up the exact number of units needed. This resulted in a significant decline in inventory levels. Once this concept was established at Toyota (ten years later) Ohno began extending it to the Toyota subcontractors.

The JIT concept has the following advantages: (1) shortened lead time (2) reduced time spent on non-process work (3) reduced inventory (4) better balance between different processes and (5) problem clarification.

Even though in today's world as bar coding has replaced the card, resulting in electronic order taking, the basic principle remains unchanged. Ohno's main goal was to develop a system for making small numbers of many different car models in order to cut waste by mobilizing a process view that extended far beyond just-in-time inventory and into the following seven areas:[5]

- Overproduction: in the case of manufacturing, this results in products that have to be discounted or stored in inventory; for service organizations, this means excess capacity and overstaffing.
- Time spent at the machine in an inefficient manner
- Shipping waste, including redundant distribution steps and delays
- Poorly designed workflows leading to processing waste and rigid procedures.
- Waste in inventory management associated with administration and backlogs.
- Excessive paperwork and "red tape", as well as wasted motion.
- Waste resulting from repair of defects as well as poor service.

The Japanese production system expert, Shigeo Shingo, who is credited with providing the detailed design for the JIT-based Toyota manufacturing system, believes that the push process used in the U.S. generated process-yield imbalances and interprocess mobilization delays.[6] On the other hand, the Japanese Kanban system of JIT pulls parts through the assembly line process, with production initiated only when the worker receives a visible cue that assembly is needed for the next step in the process. Because the process stops when a non-quality part is produced, the workers become their own inspectors and are cross-trained for a number of tasks, with the system being continuously fine tuned.

In today's world of logistics management, much of the downsizing, outsourcing, and reengineering stems from the enormous storehouses of waste that have built up over decades, resulting in overstaffing, excess management layers, complexity and bureaucracy, and lack of worker self-control and autonomy. In order to benefit from JIT, however, an organization must be willing to invest years in the careful attention to the details of its processes. Toyota found that it took about ten years and U.S. firms that have matched its success have required an equivalent amount of time.

Implementing JIT

Implementation of JIT requires a painstakingly careful attention to quality, both in purchasing and in production. Because lot sizes are usually small and there is very little safety stock to back up defective items, any quality problem severely disrupts the production line flow of materials throughout the plant. Conversely, a Macrologistics Management philosophy focuses on the fact that there must be a continuous, intensive effort on the feedback of quality issues that is a natural result of having a JIT measurement system in place that is mobilizing the logistics system to work together.

Price Waterhouse has developed a "JIT Factory Walk-Through" process that uses a formal, visual method of JIT inspection to identify areas where JIT methods can have the most impact. Their walk through facility tour, using a video camera, is a simple way of identifying areas of opportunities, especially to those who may be unfamiliar with the facility. This approach takes about a half day to complete and is based upon the following four areas of questions.[7] By going out on the floor and looking at these specific items, managers can easily identify several inefficiencies which will contribute to the improved performance of the facility, once addressed.

1. **Quality Assurance and Control:** are less than 100% of products and services visually inspected on the production line? Does conformance measurement of products occur at each stage? Is it possible for defects to move beyond the origination point before discovery? Any negative answers are in opposition to sound JIT principles and suggest aspects to look into to improve productivity.[8]
2. **Internal Transportation:** How much distance does work, materials and completed goods move about throughout the factory? Are changes in equipment layout possible in order to reduce longer distances and eliminate cumbersome and expensive transportation methods, such as fork lifts?

3. **Batch Size:** Are inventories between steps in fabrication and assembly large? Are operating parameters synchronized? Is the production lead time sufficient? If it is too long, look at the stages which take the most time and those experiencing the same kinds of problems.
4. **Worker Activity:** Is there lack of standardization among individual workers? Are the location and types of tools adequate? Is there idle time between tasks or for machine setups or during the machine cycle? Is there an imbalance between workloads of different workers or jobs?

The Move to JIT II™

While JIT is concerned with inventory, JIT II™ is concerned with using time. "This is heady stuff, and not for all organizations." These are the words of Lance Dixon, the father of JIT II™. This is the ultimate partnership program for compatible customers and suppliers, because it is the next logical step in the application of the management cycle to the value chain, through the management of time within the supply chain. It represents the use of alignment and mobilization strategies with suppliers using in-plant vendor reps to achieve breakthrough changes.

As such, JIT II™ is a major catalyst for change. While JIT is a response to competition and automation, it will not produce the necessary breakthroughs or even generate major organizational transformation. To do so requires mobilization using JIT II™, which is a key component of the macrologistics management model and is based upon the following concepts:[9]

- JIT is a worldwide recognized Japanese business technique.
- JIT II™ is a registered service mark of Bose Corporation.
- JIT eliminates inventory and the supplier and customer work closely together.
- JIT II™ eliminates the "salesman" and the "buyer", while the supplier and customer work very closely together, thus saving time.[9]

The concept of JIT II™ was developed by the Bose Corp. as a time-saving, cost-cutting approach to partnering with suppliers using the concept of concurrent engineering at its source. It is based upon a mutual-trust relationship where the supplier representative is empowered to use the company's purchase orders to place orders, which in theory replaces the purchaser and the supplier's salesperson. In practice, the supplier rep is brought into the plant on a full-time basis. This person is allowed to attend any product design meetings for his/her product and has full access to all relevant facilities, personnel, and data. Purchasing staff is freed-up from paperwork and administrative tasks, allowing them to cultivate other skills such as negotiating and sourcing. Purchase-only placement and communication is improved, time is saved, material cost reduction is realized.

The benefits are substantial for both the customer and the supplier. JIT II™ provides a natural foundation for electronic data interchange (EDI), effective paperwork, and administrative savings. Material costs are reduced on an ongoing basis. Supplier pertsonnel work on site with you and perform various planning and buying duties as well. Because supplier personnel interface daily, increased insight leads to fewer schedule change surprises. This results in reduced inventory as the supplier plans directly from the customers MRP system on a "real-time" basis. Most remaining time is spent working with your design engineering staff, thus maximizing the opportunities of concurrent engineering and cost reduction. According to Lance Dixon:

> "Lower cost is the direct result of concurrent engineering initiated very early in the design cycle. The corollary is that it is rare for competitive bidding to avchieve the cost level that can result when, for weeks or months, your project engineers have daily insights into a supplier's materials, process, and tooling issues.
>
> JIT II™ makes an old negative — 'backdoor selling' — into a positive. The companies selling directly into design engineering have been selected jointly by purchasing and engineering management."

JIT II™ brings considerable technical knowledge and support on-site, involving purchasing to design engineering. Within purchasing, supplier in-plant personnel can be seen as additional staff to address the project workload. Supplier in-plant representatives are empowered with the combined authority of the material planner, buyer, and supplier, resulting in a uniquely effective and empowered support role.

Another advantage of JIT II™ to the supplier is that they usually get an "evergreen contract", which means no end dates and no rebidding. Coupled with the EDI links and information technology exchanges which are a part of the overall logistics package, the JIT II™ concept can offer a supplier a significant strategic advantage.

The program works because it allows and encourages vendors to innovate. "It allows us to manage the account, instead of reacting to it," says Mark Finnerty, a field rep for W.N. Proctor, a Boston-based cargo agent, customhouse broker, and international freight agent. For example, a last-minute flight cancellation complicated the scheduled delivery of a shipment from Framingham to Ireland. But because Finnerty was at the Bose office, he had a quick feel for the urgency of the situation and was able to arrange for a special pickup in time to connect with alternative transportation. "Sometimes an hour or ninety minute period is the difference between making that alternative or not making it," says Finnerty. "JIT II™ allows me to sit in a customer's office workplace and get the feel of just what it is to be a shipper. I take insights and relate them to the operational people at Proctor and say these are exactly what the customer's needs are."[10] The single greatest accomplishment of JIT II™ is the efficiency in scheduling, potential cycle time reduction, and cost control that it brings to the relationship between the vendor and customer. By reducing the number of people involved, there are efficiencies for both parties — fewer purchasing personnel, more favorable pricing, reduced inventory and cycle time, and better products and processes.

Supplier Quality Management: The Missing Link[11]

Many companies leave their suppliers out of the loop and treat them as outsiders, points out *Supplier Quality* author Ricardo Fernandez. They consider their suppliers to be no more than servants who must meet their requirements as stated, or else. Accordingly, they underutilize the creative talents of their suppliers by alienating them.

World-Class companies know that the quality of their products and services is directly related to the quality systems employed by their suppliers, thus making them one of the most critical links to profit, market share and survival. When the supplier is not totally integrated into the enterprise's system of objectives, the chances of success are drastically diminished.[12] Thus, a better understanding of customer as well as supplier expectations can be achieved through a supplier certification plan, which should be based upon some "scientific" means of supplier selection.

Developing Partnerships with Partners

Dr. Kaoru Ishikawa, one of the pioneers in the Japanese Quality Movement, proposed ten principles for developing long-term partnerships with suppliers. These principles were further expanded by Fernandez[13] as follows:

1. **Both customer and supplier are fully responsible for the application of quality control with mutual understanding and cooperation between their quality control systems.**
   Each party in the relationship must be responsible for its own quality control system. Both systems, however, must interact with trust and cooperation to develop an understanding of each other's requirements. Both parties must use the same measurement techniques (e.g., statistical process control, process capability studies, etc.); otherwise, measurements of quality for the same materials or services may differ.
2. **Both customer and supplier should be independent of each other and esteem the independence of the other party.**
   This principle expands on the more general value of "live and let live" discussed earlier.
3. **The customer is responsible for bringing clear and adequate information and requirements to the supplier so that the supplier can know precisely what he should manufacture.**
   This is the "two hats" principle. Recall, the importance of a customer understanding his or her role as a supplier of information to the supplier of the goods or services being purchased so that the correct goods or services may be delivered.
4. **Both customer and supplier, before entering into business, should conclude a rational contract in respect to quality, price, delivery terms, and method of payment.**
   A contract should be agreed upon prior to conducting business. If this is to be a long-term agreement, then t should be stated as such, so that business can be conducted based on this premise. The conduct of the two parties would likely differ under this premise as opposed to a one-time purchase; therefore, this needs to be made clear. In addition, the more general commercial terms such as price, quantity, delivery, and payment terms must be specified.
5. **Supplier is responsible for the assurance of quality that will give satisfaction to the customer and is responsible for submitting necessary and actual data upon customer's request.**
   In addition to responsibility for the quality system at his or her site, the supplier is also responsible for meeting or exceeding the quality levels agreed upon with the customer, thereby satisfying the customer. In order to build trust and show proof of quality to the customer, the supplier should respond to the customer's request for the actual data on key quality characteristics of the product or service. This could be in the form of control charts, capability studies, reliability studies, etc.

6. **Both supplier and customer should decide on the evaluation method of various items beforehand, which will be admitted as satisfactory to both parties.**
   When evaluating quality levels, a good operational definition of the quality characteristics must be determined beforehand. This must be negotiated between the supplier and the customer so that there is no ambiguity in the expectations of either party. If these characteristics are well defined, there will be little ground for future disputes and a stronger relationship can be built.
7. **Both customer and supplier should establish in their contract the systems and procedures through which they can reach amicable settlement of disputes whenever any problems occur.**
   Disputes can easily destroy a long-term customer-supplier relationship, especially if no agreement has been reached on the methods to be used to resolve these disputes. For example, today many contracts refer disputes to binding arbitration. This keeps disputes more amicable and makes them less costly to resolve for all involved. Other methods could be utilized, such as predetermined, mutually agreed upon outcomes for different standard types of disputes.
8. **Both customer and supplier, taking into consideration the other party's standing, should exchange information necessary to carry out better quality control.**
   This principle is similar to Principles 1, 3 and 5. It is more general and explains the need for good communication and the exchange of information in order to perform better quality control.
9. **Both customer and supplier should always perform control business activities sufficiently, such as ordering, production and inventory planning, clerical work, and systems, so that their relationship is maintained on an amicable and satisfactory basis.**
   Continuity and repetition are key to this principle. Basically, a good long-term relationship benefits from frequent contact; therefore, it is of mutual interest that both parties conduct many transactions on a continuous basis. This promotes trust and facilitates understanding of each other's systems and processes.
10. **Both customer and supplier, when dealing with business transactions, should always take full account of the ultimate consumer's interest.**
    Last, but certainly not least, is the need to keep the ultimate consumer in mind. Without that ultimate consumer's purchases, neither party can survive. When making decisions in this relationship, both parties

must maintain the highest emphasis on the needs of that ultimate consumer. For example, a very high-quality product with a very high price may not be in the best interest of the ultimate consumer who wants value for his or her money. The ultimate consumer believes that the manufacturer or service provider is being as efficient as possible to provide the best value for the money.

Steps in Supplier Certification

There are five steps involved in supplier certification:

1. **Verify** the capability of the supplier in meeting the needs of their customers in all areas. This can be accomplished through the supplier audit. The buying organization sets the criteria it wants to verify that the supplier meets. Some of the criteria should be standard for all suppliers, while others are more commodity or industry specific.
2. **Motivate** suppliers to continue to improve their processes and the resulting products and services through a series of incentives for attaining the higher levels of certification. Typically, there are three levels: (1) Quality Vendor (2) Certified Vendor and (3) Excellent Vendor.
3. **Improve** key supplier processes in the value chain. This is accomplished through Process Mapping and benefits both the supplier and the customer organizations, along with establishing a win-win relationship.
4. **Assess** continuously the supplier's capabilities. This needs to be done to ensure that the supplier who attains a level of certification maintains that level and is able to try to move to the next one.

Supplier Quality Management (SQM) Implementation

There are a dozen or so activities involved in implementation:

- Assess external customer needs
- Assess internal, stakeholder needs and develop a plan
- Design SQM system structure
- Provide plenty of training
- Implement Supplier Policy Deployment
- Hold supplier symposia

- Perform planning for supplier quality projects
- Provide Logistics Improvement support
- Train suppliers in the Macrologistics Management system
- Create a system for continuous monitoring
- Perform supplier certifications and review prior certifications
- Review and revise the Supplier Quality Management system

One of Dr. Deming's Fourteen Points is to accomplish the transformation. A well-mobilized Supplier Quality Management system will go a long way in helping an organization to logistically partner with their key suppliers and vendors, especially when the principles of JIT II™ and Adaptive Learning are used as part of the effort to transform and reduce the supplier base.[14]

Summary

JIT II™ has emerged as a cost-saving and cost-sharing way to align with suppliers. If the in-plant is empowered and supported by Information Systems infrastructure, major breakthroughs are possible. At Bose, a competitive posture with respect to Japanese electronics companies has been facilitated by JIT and JIT II™ cost savings, reduced cycle time and concurrent engineering modifications.

These strategies have helped Bose and other organizations position products competitively and effectively mobilize their strategies in a highly competitive marketplaceas as shown in Profile 7.1. Without global logistics approaches, organizations such as Bose cannot be competitive and prosper. These companies have prospered through extremely lean years in very competitive industries. The JIT II™ process is now seen as a key tool for market share expansion to be used for future sales and marketing growth.

References

1. Portions of this chapter were based upon *"Applied Operations Management,"* James Evans et al., West Publishing, St. Paul, MN, 1991, pp. 706–714.
2. Richard Schonberger, Japanese Manufacturing Techniques, *The Free Press,* 1982, pp. 17–18.
3. Masaaki Imai, *Kaizen,* The Kaizen Institute, 1986, pp. 89–90.
4. Automation in a small area requires JIT. In Japan, there was no space for backparts to be stored. Thus, automation was used hand in hand with JIT to increase plant output in the same area. Also, they reduced stages as they automated resulting in a streamlined production cycle, made possible by JIT.

5. Peter G. W. Keen and Ellen M. Knapp, *Business Processes*, Harvard Business School Press, 1996.
6. Joel Ross, *Total Quality Management*, St. Lucie Press, 2nd edition, 1995, pp. 165.
7. Mark R. Jamrog, "Just-In-Time" Manufacturing: Just in Time for U.S. Manufacturers, *Price Waterhouse Review* #1, 1995 (orig. 1988), pp. 27–28.
8. Alexander Hiam, *The Vest Pocket CEO*, Prentice Hall, 1990, pp. 116–117.
9. JIT II™: The Ultimate Customer/Supplier Partnership Program, *Purchasing Magazine*, September 1991.
10. The Ultimate Customer-Supplier Relationship at Bose, Honeywell, and AT&T, by Martin Stein, *National Productivity Review*, Autumn 1993, pp. 543–548.
11. The Supplier Quality Management model presented here is based upon the book *Total Quality in Purchasing & Supplier Management*, by Ricardo Fernandez, St. Lucie Press, 1995.
12. Ibid. pp. 6–7.
13. Ibid. pp. 70–73.
14. The Supplier Quality Management portion of this chapte was adapted from a work by Rick Fernandez called *TQ in Purchasing/Quality Management*, St. Lucie Press, 1995.

Bose Uses JIT II™ for Competitive Advantage 7.1

Bose is a company on a mission to provide outstanding sound experience to everyone in the whole world. It is a privately held corporation that is the world's largest manufacturer of component-quality speakers. Its headquarters is located 23 miles west of Boston. Bose has three manufacturing facilities located in Westboro, Massachusetts; Ste. Marie, Quebec; and Carrickmacross, Ireland; with planned facilities in Hillsdale and San Luis, Mexico.

Speakers are the most competitive of the components of the audio business, with dozens of manufacturers in the United States, Europe and the Far East. A diverse array of designs and technologies revolve around three critical subassemblies to all speakers:

- **Transducer:** The traditional speaker had at least two "woofers" that are low-frequency bass sounds and "tweeters" that are high frequency.
- **Electronics:** Featuring increasingly more sophisticated integrated circuit boards that managed amplification and maintained system balance.
- **Cabinet:** The role of soundwave director in an attractive suit was understood only too well by Bose. The exterior surface on the cabinet had to be perfect in all aspects.

Bose is an excellent example of an organization using a combination of Macrologistics strategies to achieve competitive advantage. Central to their strategy is the use of JIT II™ to fully mobilize their logistics operations. Information Systems provided the interconnectivity, while their use of Process Management helped them redefine the marketplace they serve. Their innovative strategies for maximizing their supply chain resulted in substantial cost reductions while shortening the delivery cycle.

Background Information

Bose Corporation was founded in 1964 by Dr. Amar Bose, a professor of Electrical Engineering and Computer Science at MIT. He befriended a man by the name of Sherwin Greenblatt, who, as the Bose Corporation's first employee, planned to build a company based upon innovation in acoustics and electronics. For three years, they focused on the "hobby side of the business." Then, in 1968, Bose launched the 901 speaker, which simulated the feelings of live sound by radiating the waves to the listener directly, as well as off walls, ceilings, etc. Over the next ten years, this resulted in a dramatic expansion of this line of speakers, which got smaller and smaller and better and better.

The decade of the 1980s saw the introduction of the Bose speaker line into the automobile industry. Cadillac was the first to use them in their top-of-the-line models in 1982, followed by Toyota, Nissan, Acura, and others. In 1990, the company embarked upon a three-pronged strategy, according to Lance Dixon, Vice President of Logistics:

> "First, the company sought out new markets around the world. By 1990, we were the highest selling manufacturer of stereo component-quality speakers in Japan, as well as in Holland, France, Australia, and other countries. Management at Bose believes that the desire for quality sound is universal and plans to continue opening new markets around the world."

During the early 1980s, Bose hired Lance Dixon to help run the procurement operations, and his background suited him well for the challenges ahead. He was an ex-marine who began his career at Honeywell Corporation, where he managed an engineering support facility that performed printing and photographic work. Responding to his suggestions of how to get more out of the purchasing dollar, management put him in charge of centralizing the purchasing of printed materials. When Honeywell bought GE's computer division, Dixon established a network of twelve warehouses to provide promotional support for the combined Honeywell/GE sales force.

After spending three years tying together the procurement process, he standardized prices with vendors and established consistent procurement practices for multiple Honeywell divisions. One of the breakthroughs for Lance Dixon was the understanding that the company produced systems as well as components and was open to new systems approaches. In the early days, good hi-fi was available only to those consumers willing to invest a lot of time, money, and patience in their systems. By 1992, however, they expected hassle-free sound. Accordingly, the marketplace for integrated audio equipment had grown to more than twice its size of the market for separate audio components. In 1987, after 14 years of development, Bose introduced the 'Acoustic Wave Music System', a completely integrated, portable, high-performance music system incorporating speakers, an AM/FM receiver and a cassette tape deck (to be followed by a digital-disk player). This meant that a radically new approach needed to be taken with the JIT process already in operation.

Frustrated by his inability to secure adequate resources to run an expanded purchasing function of the future, Dixon developed a 'profit-center approach' where he was allowed to reinvest half of the 'savings' he achieved below standard cost. He also instituted a program to pay cash incentives to buyers for savings achieved on purchased items. The program was so successful that in 1990 Dixon proposed to change the relationship between Bose and certain suppliers under a program he dubbed 'JIT II™.' In JIT II™, the vendor replaced the salesperson, the buyer, and the materials planner. These 'reps' were stationed full time at Bose but were paid for by the vendor. According to Dixon:

> "I told management that I could get the people I needed to get the job done and not add anything to the Payroll! Well, this really got some attention. I told them about my plan to locate JIT II™ in-plant supplier representatives integrated with the various Purchasing sections. The response was predictable and I was asked 'It seems like putting the fox in charge of the hen-house, does it not?' by a senior executive. My response was the same then as it is now, JIT II™ actually improves the controls and builds a collaborative work environment. The supplier has a legal, bonded responsibility to Bose and pilferage and loss will be reduced, not heightened. If we were to find problems of mismanagement or corruption, and we never have, our legal remedies are much cleaner and swifter in judgment. The JIT II™ rep works for the supplier and, once trust is established, the results and rewards of shorter cycle time and reduced costs are enormous. The following JIT II™ suppliers were found at Bose:

- Doranco — in mechanical New Product Purchasing.
- G&F Plastics — in Corporate Purchasing and Bose plant locations.
- United Printing — in MRO Purchasing and plant locations.
- Monroe Stationers — in MRO Purchasing.
- American Presidents Lines — in Transportation.
- Roadway Express — in Transportation.
- Proctor Inc. Import — in Transportation."

Until 1988, no purchasing at Bose had been done by the plants; instead, all items had been purchased by the corporate procurement department. By 1990, much had been decentralized, and the plants did their own day-to-day purchasing, typically in line with contracts that were negotiated by Dixon's central unit. A typical plant spent about $100,000 million per year on items purchased from an active base of about 200 vendors. Westboro spent about $140 million per year on items purchased from an active base of about 200 vendors. About 50% of the plant's purchasing dollars were spent in five categories: electronic components, plastics, printing, corrugated boxes/packaging, and cables/cords. Purchasing was planned in a three-stage cycle, as outlined by Dixon:

- **Stage I:** Business planning. The marketing department at Bose Corporation prepared multi-year plans.
- **Stage II:** Aggregate production planning. Based on the business plan, Westboro prepared a production plan that specified the capacity, tooling, and material volumes that would be needed over the next one to two years.
- **Stage III:** Production scheduling. Based on the aggregate production plan, schedulers at Westboro prepared a detailed "master schedule" outlining requirements for capacity, personnel, and material over the coming 12 months. Production of earlier months was scheduled at a greater level of disaggregation than for later months.

The Westboro materials manager coordinated scheduling, purchasing, and inventory. Five people reported directly to him: an inventory manager, a warehouse manager, and three materials managers. The inventory manager was responsible for tracking and managing overall inventory levels at Westboro, while the warehouse manager oversaw the operation of the plant warehouse. Each materials manager performed the planning and purchasing to support one production line and was assisted by a production control supervisor, a master scheduler, and a purchasing supervisor.

Inventory Problems

Dixon describes the overall inventory situation at Bose in this manner:

> "Purchasing supervisors supervised a group of buyers who procured all materials for one production line. Buyers were responsible for managing quality, cost, and delivery. Unlike Corporate Procurement, most buyers at Westboro were not engineers and instead had come up through the ranks as administrators or expediters. Buyers at Westboro typically started on easier commodities, such as hardware or operating supplies, and then moved on to more difficult categories such as plastics and electronics.
>
> Most of the buyers' time at Westboro was taken up by inventory planning, which encompassed three activities: deciding what to order, placing new orders with vendors, and adjusting delivery schedules to accelerate or delay delivery on ordered parts. Another 15% of buyers' time was spent on revisions to existing parts; usually this entailed updating documents or ensuring that revised parts met quality levels. The remaining 10% of buyers' time was devoted to renegotiating contracts with existing vendors or, occasionally, switching to new vendors".

Westboro buyers preferred vendors who maintained a secure financial position, were located close to Westboro, could provide fast delivery, maintained consistent production processes (as measured by the use of statistical process control and a quality rating system) and provided good references through Corporate Procurement or other customers. The average lead time on purchase orders placed by Bose Corporation was four to six weeks, but one-third of all purchase orders had less than 10 days' lead time. About 35% of all orders were changed within 30 days after placement.

Dixon's Vision

Dixon's vision was to combine partnering with information systems and concurrent engineering in a three-legged stool fashion. A great deal of his philosophy centered around the "Partnering Principle:"

> "People have bought into the idea of partnering, they just don't know how to implement it. It's a philosophy in search of substance. Most people will tell you, 'Yes, I love motherhood and apple pie and partnering.' But when you ask them what they mean and have done, you generally get an awkward silence. JIT II™ brings our

vendors into the company — literally. The vendor places purchase orders on himself. Therefore, he is empowered by Bose. It's like a blank check, but with the proper controls and monitoring at buyer level to see that the partnership (alliance) endures and prospers. In fact, Dixon claims, it was actually catalytic in that it changes your perspective and allows you to realize unforeseen benefits".

As a starting place, the commodity area of plastics and printing were the initial candidates. Bose spends about $15 million on purchases of plastics components, which originally involved five vendors having 60% of the volume. Bose chose G&F Industries, a 60-person, $12 million operation, as their JIT II™ plastics vendor. On the printing side, Bose chose United Printing as their JIT II™ supplier of instruction booklets, warranty cards, and promotional materials. They then expanded the concept into other areas such as transportation, metals, and export/import.

The Transportation Cycle at Bose

Transportation is divided into two categories: boat/plane and truck. W.N. Proctor was chosen as the JIT II™ in-plant for the boat/plane area. The trucking was staffed by a P-I-E representative until their bankruptcy, when Roadway became the JIT II™ vendor. Like all Bose in-plant personnel, these reps were free to follow a career path at either the vendor company or at Bose. The reps' best weapon for fighting glitches in the movement of goods and materials was a computer system that tracked the merchandise and allowed them to clear goods with customs before leaving the ship or plane. The result was that more than 50% of Bose's inbound freight enjoyed paperless clearance through the U.S. Customs, putting them in the top 1% of all companies, according to American Shipper magazine.

JIT II™ and EDI at Bose

When Dixon first implemented JIT II™, he also wanted to include a less-than-truckload carrier. Since he wanted EDI links, he looked around for someone who had the best EDI available. This turned out to be P-I-E, and they were selected despite their unknown track record with Bose. The system worked fine, and when P-I-E went out of business, Bose again selected a vendor with EDI capability — Roadway. When there was a problem with a delivery, the in-house rep fixed the situation in a matter of minutes (literally) instead of following a convoluted process in which the consignee called Bose, who called the vendor, who investigated and called back, and so on. Dixon describes the system in the following manner:

> "This is the logistics transportation version of 'one stop shopping,' and even allows Bose people to examine damaged materials using the Roadway terminals in order to target a specific problem area (instead of berating Roadway in a general way). Bose has been known to visit a warehouse to perform a management visit, especially if it was experiencing a heavy loss due to theft. The idea of the connected information systems allowed people to swing into action before small problems became bigger ones. Future plans call for applying more EDI links on the outbound tracking operations as well, but to date many of the shipping lines are behind the times when it comes to automation, although some are currently changing, e.g., both American President Lines and Sea-Land have integrated the application of JIT II™ to their operations."

Summary

In the future world of Macrologistics Management at Bose, there is a natural evolution in the products produced by each plant. Initially, plants only produced components for other plants, but then became increasingly self-sufficient. Over time, plants are expected to expand their range of manufacturing capabilities, integrating forward to produce finished products and backward to produce as many of the components required for these products as possible. Lance Dixon describes controlled integration as follows:

> "If Bose were big enough, we'd do everything ourselves. Obviously, there are some areas we'd never integrate into, such as owning the steamships or the trucking lines that transport our speakers. However, if it's fundamental to the quality or performance of Bose products, then we want to control it."

Because of this natural desire to control the integration when pursuing an expanded range of operations, management has cited a few potential problem areas. One is that the vendor and Bose have different priorities and agendas, which could lead to misfits if not carefully monitored. Also, relying upon the vendor might preclude Bose from developing internal capabilities in manufacturing parts expertise. Finally, vendors can never understand Bose operations and needs as well as company employees. Overall, though, management is very excited about JIT II™ and has helped a number of other companies understand and begin to integrate this key concept of Macrologistics Management for themselves.

The concept works well in either sole source or multiple vendor per commodity situations. For the most part, Bose only implements one JIT II™ vendor in a given commodity. Dixon summarizes this point as follows:

"We have found that it is not a problem to have one 'most-favored nation' status vendor relationship, and still carry on a professional, mutually satisfying, fair relationship with other competing vendors in the same commodity. We use the following criteria to detect the possible vendors:

- An excellent vendor: with volume over $1 million
- Current goods quality: based upon a substantial number of purchase order transactions
- Current goods delivery: using evolving technology, but not at a revolutionary change pace
- Current goods cost levels and engineering support: and not be involved in company non-trade secret or sensitive technology area

Our JIT II™ vendors are using JIT II™ as a sales tool and have implemented it with other customers, with our guidance. This concept provides a built-in competitive advantage and is presently being licensed to a number of large organizations throughout the U.S.

For Bose to be able to continue to grow at 25% per year compounded rate, bridging and teamwork will be crucial. If the retailers and Bose collaborate, they might be able to cut one level and reduce inventory for the mutual benefit. Supply chain simplification requires information sharing and joint planning. But the end result can be substantial reductions in the cycle time due to the flexibility of early ordering and the system interconnections worldwide. We believe this puts U.S. on the threshold of redefining cycle time in a breakthrough manner."

In answer to queries about the acceptance of the concept in the manufacturing community, Dixon proudly points to the fact that Bose has had to create a "Visitor-Orientation" day once a month, where a special JIT II™ team conducts seminars and tours for executive from around the world. Concludes Dixon:

"Up until a few years ago, the company was content to give this concept away as a matter of good public relations. Our approach now is to offer the services on a break-even basis, with any remaining funds going to provide scholarships for needy student individuals. In recent years, it has been the cornerstone of the Bose sound visionexperiencetoeveryonein the whole world',the world of JIT II™."

A Day In The Life: JIT II™ Reps in the Transportation Process 7.2

A typical day for our in-plant JIT II™ Rep Penny Archer, is quite different. Quite a bit of it is spent controlling movement of freight by truck, sea and air on a world-wide basis. Let's follow Penny as she interacts with the Roadway Express and the Proctor mainframes on an international order she is expediting.

She enters both systems through terminals provided by these "JIT II™ partners" and finds what she is looking for. (Last week she was stymied by a problem, which the part-time vendor system support professional was able to fix quickly). As she passed the Stats Control Room, she couldn't help but notice the latest measure: an excess of 50% paperless EDI clearance through customs for the seventh month in a row. It sure feels good to be #1 in that category and in the top 1% for the year. What's our JIT II™ secret? We link our computer systems with the Customs computer systems! We call this Breakthrough Thinking at Bose.

Next, it's 11 a.m. and time to check on the Bose container leaving on a boat from Kaoshung, Taiwan and the connected package on a plane from Japan. This info is available in a "common record" profile format by the location freight forwarder and is easily accessible. The profile contains every aspect of the shipment and is forwarded to and from the Boston location of Proctor, Inc. This data is also made available to all Bose terminals as needed.

At 2 p.m. another visit was paid by Penny to the interconnected mainframe network. Boy, the fingertip availability of that data is so powerful, it changes the way we do business. Because it is real-time, we have daily opportunities to redefine cycle time. When a day or two is the difference between impacting a Bose factory production schedule, the Transportation rep provides Bose Procurement with the extra margin to avoid the roadblock or problem.

It's 3 p.m., and here comes the boss. (I wonder if he noticed the progress we've been making with JIT II™?). "Good morning, Penny. Your good work of 'fine tuning' between our Procurement people, Transportation, and our plant engineers made a difference last month once again! We saw an immediate material cost reduction. Our communication process is now two-step instead of the four. Penny couldn't help smile at his reference to the 'four horsemen' of logistics: planner-to-buyer-to-salesman-to-plantperson."

Well, it's 4 p.m. and almost time for Penny to call it a day. It sure is nice to know you made a difference with the JIT II™ system once again. And she can't wait for tomorrow, for she has a meeting on the Living, Breathing Standards program and will be sitting in on an early design meeting with the engineers and others. So, let's follow Penny to visit with Cliff LaBonte, the In-Plant Rep from G&F Industries, and contrast his workday to Penny's (Cliff is getting ready for his trip to Mexico and is just getting ready to leave, so we only have time for a quick visit).

By the way, G&F Industries supplies plastic injection molding tooling, plastic parts and metal parts to Bose, shipping to various Bose plants worldwide, or utilizing the G&F Ireland facility.

Chris starts some days at his plant in Sturbridge, MA, where he controls various production schedules with a status review of Bose in-process parts. He arrives at the Bose Westboro manufacturing plant where he confers with John Argitis, Jr., the other G&F In-Plant. John is heavily involved in the daily planning and ordering of G&F material for this particular Bose plant, using Bose purchase orders, from his location in the plant purchasing department. (G&F is one of two companies who have asked to place two persons to keep up with all aspects of the business.)

At 11 a.m., Chris arrives at his office in the Corporate Purchasing Department where he confers with the Bose Framingham plant material planners and receives related material requisitions. He "calls them in" to his plant after sign off by Bose Purchasing Manager on orders that exceed his dollar authorization per order, the same as any Bose buyer. At 1 p.m., he attends a Bose New Product Project review at the Bose Mountain Headquarters location, gathering information of any importance on parts G&F will be supplying. 2 p.m. brings Chris into contact with Bose design engineers who have questions on process possibilities and cost tradeoffs on a plastic part and various materials. At 3 p.m., a quality control issue is addressed with Corporate & Plant Quality personnel.

An oft-cited axiom about researchers is that the top 10% produce most of the results. Bell Labs decided to find out why. A team of consultants examined the behavior of the lab's top performers and compared them to that of average, less productive performers. They identified nine work strategies that made the top performers successful.

1. **Taking initiative:** accepting responsibility above and beyond your stated job, volunteering for additional activities, and promoting new ideas.
2. **Networking:** getting direct and immediate access to coworkers with technical expertise and sharing your own knowledge with those who need it.

3. **Self-management:** regulating your own work commitments, time, performance level, and career growth.
4. **Teamwork effectiveness:** assuming joint responsibility for work activities, coordinating efforts, and accomplishing shared goals with coworkers.
5. **Leadership:** formulating, stating, and building consensus on common goals and working to accomplish them.
6. **Followership:** helping the leader accomplish the organization's goals and thinking for yourself rather than relying solely on managerial direction.
7. **Perspective:** seeing your job in its larger context and taking on other viewpoints such as those of the customer, manager, and work team.
8. **Show-and-tell:** presenting your ideas persuasively in written and oral form.
9. **Organization savvy:** navigating the competing interests in an organization, be they individual or group, to promote cooperation, address conflicts, and get things done.

Of these nine strategies, taking initiative was viewed as the most important. Six other skills — self-management, perspective, followership, teamwork effectiveness, leadership, and networking — were next in importance. Organization savvy and show-and-tell were considered icing on the cake.

Source: Robert Kelly and Janet Caplan, "How Bell Labs Creates Star Performers," *Harvard Business Review* (July-August 1993), pp. 128–139.

JIT and the Management Accountant 7.4

A noble philosophy, this JIT, or as author Ian Cobb calls it: "Just Intelligent Thinking." Eliminate waste from all parts of the manufacturing process through the use of the Continuous Improvement Cycle is the author's basic definition of Just In Time (JIT). Like all things noble in England, however, JIT has been subjected to intense scrutiny and has been met with some skepticism. These are a few of the things discovered in a recent survey of 211 senior Chartered Institute of Management Accountants (CIMA) members from 205 companies across the United Kingdom. The survey based its questions upon E. J. Hays' seven aspects of JIT. However, the study is probably more accurate to say that these seven aspects are the engine or mechanism of JIT rather than the "seven immovable pillars" of JIT.

The first aspect, JIT purchasing, requires smaller, more frequent delivery of raw materials. The second, machine cells, are where workers and machines are grouped in order of production. The third is set up time reduction, hopefully eliminating time-consuming set ups. The fourth aspect is uniform loading, which matches production to the rate at which the product is needed by the customer (internal and external). The fifth is the use of a pull system or *kanban* which eliminates backstock. The sixth aspect incorporates logistics value stream quality to eliminate inspection. The seventh is employee involvement at all levels of the company.

These elements allow for a wide range of applications, and as seen by Cobb in the survey, most companies have embraced at least some of the aspects of JIT. The survey pointed out a lack of understanding of some of these aspects and underscored the need for better education of future CIMA members in the aspects and applications of JIT. Many benefits of introducing JI were seen by Cobb in the survey, such as reduction of accounting transactions, increased manufacturing flexibility, improved customer service. However, reasons for not introducing JIT, such as lack of knowledge about JIT and reluctance to face challenges posed by JIT point out the need for better education. Overall, 33 management accounting system companies reported a total of 50 changes in the areas of accounting for work in process, direct labor, and manufacturing overhead. As JIT is worked into more companies every day, one thing is clear: the management accounting systems of the past are going to change and the criteria for evaluating management accountants will change with them.

Source: Ian Cobb, Management Accounting, London (Mac), Vol. 70, Issue 2, Feb. 1992, pp 42–44.

Inventory: Do Your Policy and Practice Stack Up? 7.5

A major factor in productivity and financial return rates of an organization in the 1990s are management policies and practices for inventory control. Attention to details in the key areas of Planning, Purchasing and Assuring Quality can yield high payoffs if management can remember to "walk the talk." The use of a Manufacturing Resource Planning (MRP) system, which demands the commitment of top management, can prove invaluable during the Planning stage.

On the other hand, Purchasing personnel face a dramatic change in the way they do business. In the past, purchasers who caught flak for previous problems overcompensated by increasing their order quantities and bringing in material early. Managers should now realize that purchasing personnel should work more closely with suppliers as team members. They will help to keep material costs down and inventories in check by ensuring high quality of incoming goods and on-time delivery.

Logistics-based Quality Management can reduce Work in Process (WIP) by encouraging the use of logistics-based quality control tools such as: design reviews and prototype testing, employee participation, vertical communication, employee training, Statistical Quality Control (SQC) methods, automation and customer feedback charts. If inventory is to be controlled, management must communicate effectively to all levels of an organization and keep in mind that some of the things they are doing can effect the company negatively and raise WIP.

Source: Ken Kivenko, *Automation (PDE)*, May 1988, pp. 51–52; Ricardo Fernandez, *Total Quality in Purchasing and Supplier Quality Management*, St. Lucie Press, Boca Raton, FL, 1995.

The Impact of Just-In-Time Inventory Systems on Small Businesses 7.6

A key emerging logistics strategy is that Just-In-Time (JIT) inventory management systems, used to reduce inventories and improve quality and productivity, are being adopted more frequently by large manufacturing corporations, thus presenting new challenges for small businesses as well. JIT, a "pull" system (assembly line triggers withdrawal of parts from preceding work centers), has two aspects that are heavily emphasized by manufacturers: Just-In-Time purchasing, and Just-In-Time delivery and transportation. JIT purchasing, which relies heavily on a dependable suppliers network, calls for manufacturers to deal with fewer suppliers, small lot sizes, and statistical quality control (SQC) techniques.

In addition to using fewer suppliers and signing larger contracts (with suppliers), JIT manufacturers are working with smaller lot sizes, thereby reducing unnecessary inventories and freeing storage areas to reduce cost and improve quality. SQC is a powerful problem-solving tool that pinpoints variations and their causes, and eliminates after-the-fact inspection — an expensive, wasteful procedure that rarely detects the causes of poor quality.

More recently, JIT manufacturers have encouraged their suppliers to form "focused factory" arrangements which enable the supplier to focus on a limited number of products and become a specialized maker to a major manufacturer. This usually leads to suppliers relocating closer to their respective manufacturer, one of the tenets of JIT delivery and transportation. As well as the elimination of centralized loading docks and staging areas, another benefit is the use of information sharing and microcomputer use exposing small businesses to computerized information systems.

The implications of JIT on small businesses are profound, such as better customer relationships and stable product demand. Not all the benefits come easy, though. For example, suppliers must institute SQC and understand logistics freight economics, but the benefits of long-term contracts, smooth production, improved quality, and reduced scrap and rework greatly enhance the potential for increased market share success of small businesses.

Source: A. T. Sadhwani and M. Sarhan, *Journal of Accountancy (JAC)*, Jan 1987, pp. 118–132; Ricardo Fernandez, *Total Quality in Purchasing and Supplier Quality Management*, St. Lucie Press, Boca Raton, FL, 1995.

8 ValueStream Quality System*

Jeffrey Vengrow and Frank Voehl

In the global marketplace, commercial Darwinism is alive and well. Survival of the fittest in this sense has little to do with genetics, but it has everything to do with developing a competetive advantage. In today's commercial setting, a well-developed competitive advantage is necessary to successfully survive in tomorrow's often hostile competitive environment. Survival is often associated with adaptation and change. What constitutes a competitive advantage today may no longer be one in the future. Since the Industrial Revolution, many organizational models have appeared reflecting an accelerated pace of commercial evolution. Organizational models may be small on complexity but large on challenge for successful implementation.

Today, there is a new enterprise model that is borne out of world-wide competition. This model is called the ValueStream Quality System. By applying *quality principles* in combination with a *process view* of an *extended enterprise*, a customized model for successful survival can be created.

Achieving commercial dominance within a selected market or region is an outcome measure of a well-developed ValueStream model. Developing the financial strength to seize opportunties coupled with the leadership skills to set both the direction and the pace of a market niche are symptomatic of these enterprises.

---

* This chapter is adapted from a book in progress titled *ValueStream Management: The New Wave Value Proposition* by Ralph Lewis, Jeffrey Vengrow, Bill Meyer, and Frank Voehl.

Stakeholder Value

Stakeholder value is a very complex and misunderstood concept. It is not simply the price a customer pays for a product vs. value received. It is that and much more. It is the perceived balance between the things people believe in exchange for what they give up in order to get them.[1]

Each stakeholder group has a special role to play in the success of the company in the ValueStream. Management's role in the transaction is to ensure that the stakeholders are satisfied. Stakeholder value is one of the most critical assets that a company has.

Evolution of the Value Stream

Little research is required to reveal how the commercial world of today is littered with the remnants of enterprises until recently believed impervious to the forces of a changing environment and the need to belabor stakeholder value.

Just glance at the evolution of commercial models starting from the days of individual craftsmanship to the industrial revolution of mass production leading to automation, vertical integration, horizontal integration, decentralization, globalization and even miniaturization.

Prior to the industrial revolution, individual craftsmanship supported small regional markets. The expansion of the enterprise was limited by the personal capacity of the owner/operator. This meant that a relatively small population was served but well-established relationships with the craftsman were a recognized part of the value. Since the products were made one at a time, the craftsman could incorporate some unique features desired by each customer. There were also a limited number of competitors unless a buyer was willing to suffer the travel. Uniqueness, community relationships, skill mastery, and transportation barriers collectively constituted a competetive advantage for the local craftsman.[2]

The industrial era brought with it the capability of producing more products faster to much larger markets. Initially in the United States, the manufacturing processes depended upon unskilled and often non-English speaking labor. To accommodate these challenges, the work was divided up into a sequence of individual and repetitive steps that were easy to explain and to learn. There was a downside to this successful mass production model. Workers were charged with following orders without input or creativity. There often lacked the perspective of how each component or step integrated with the finished product. Quality in the worker's eye was based on that incremental step, not on the completed product. In addition, relationships with the customers were severed from those who made the product. This represented a significant departure from the master craftsman who had a real knowledge of the customer's needs and expectations. Unlike the factory worker of the 19th century, the craftsman had an understanding about how each part of the product was connected to the one before it and the one after it based on a complete knowledge of the whole. The factory owner might have been the entrepreneur, but the largesse of the enterprise prevented that focus on the customer from permeating the entire organization. Today, an organization which adopted this model would be so disconnected from itself that it would quickly be overtaken by even a low-productivity enterprise which communicated well among its the entire ValueStream.

The expanding domestic market had an insatiable appetite for goods which created opportunities for many businesses to participate without serious competition from each other. Labor was cheap and plentiful. The addition of railroad transport made national distribution and logistics both possible and cost effective. Availability of products at affordable prices was more important to consumers than high quality or value added services. These factors coupled with an isolationist mindset in government represented a competitive advantage for domestic industry. Today, protective tariffs are under siege as global sourcing of goods and services multiplies, and the skilled labor required to provide them is both limited and expensive. By integrating the resource pool available within the ValueStream Quality System, resources can be utilized more productively to offset today's demand for value at competitive prices.

Refocus on Manufacturing Adding Value

Focus on Quality in manufacturing became popular in the 1980s as America's core industries lost market share to foreign competition, primarily from Germany and Japan. While their economies had recovered from the war and much of their industrial infrastructure was new, U.S. manufacturers had become complacent during the robust era following the war. In addition, Japan in particular began to link their manufacturing resources with suppliers in a way never considered in the U.S. By developing structural alliances, the Japanese were able to integrate their logistics requirements with key suppliers. This kind of extended enterprise or Kierestu was considered illegal in the U.S. as anticompetetive and monopolistic in nature. For Japanese businesses however, this reduced operational costs associated with Just-in-Time deliveries, lower inventory deliveries and less material handling for the primary manufacturer (see Profile 8.5). These methods proved to be a competitive weapon deployed most effectively to compete with U.S. manufacturers. This model proved far from perfect because this extended enterprise was not integrated, only delegated. Therefore, while Japanese manufacturers were touting lifetime employment and less material handling, their suppliers were carrying inventory for them and routinely laying off and recalling their own employees in order to make this possible. Clearly this model represented an improvement over American business practices in that era, but it lacked a balanced whole systems approach which is a more effective model for today.

The emergence of higher quality, lower cost foreign goods with more features and options accelerated over a decade. The cry for protective legislation and the pressure to "buy American" failed to quell the rising tide of foreign product introduction. Quality was adopted as a significant part of a strategy to restore our competetive edge. Reduction in process variation within manufacturing plants reduced operational costs while improving product reliability. In addition, global sourcing and manufacturing evolved as a concurrent strategy to take the offensive to foreign competition while taking advantage of lower cost opportunities. By producing closer to either the raw material source or the user markets, U.S. businesses developed materials distribution as a competitive weapon. Without a clear and organization-wide system of quality, however, this winning strategy proved only marginally successful upon implementation despite low-cost labor, government support and minimal regulation. The lack of an integrated quality system prevented U.S. businesses from maximizing a global operational strategy. International cultural and work practices were not sufficiently substituted by quality systems and this substantially impacted on results. The failures in quality, efficiency, and dependability reflected the absence of quality systems within our industrial infrastructure.

The manufacturing processes of U.S. businesses resumed the tightest quality scrutiny not seen since the days of our World War II defense industry. As the 1990s began, the quality notion was firmly entrenched in production departments and later extended to other functional parts of the organization.

Competitive Advantage Lost

The reduction in operating costs realized through improvements in quality of manufacturing succeeded in preserving and even regaining some lost competetive advantage, but only temporarily. Why?

First, many of the quality improvements included automation, robotics, and other forms of dedicated process technology. While these manufacturing strategies were designed to both reduce costs and improve quality based upon consistency, this capital intensive approach increased the cost of customization precisely when customers had developed a penchant for options. In addition, this technology required higher levels of skill to maintain, troubleshoot, and modify the equipment at a time when demand for these skills exceed their supply.

Second, quality in manufacturing is becoming more common. Unless a method to turbocharge the impact of quality principles is established, today's model of quality production will no longer be a competitive distinction by itself.

Third, the structural separation between producers and customers continued restricting a broad based understanding of customer satisfaction criteria. A key part of the failure of the mass production model of the industrial revolution was that process improvements, and design integration did not involve those directly involved in creating the added value. Workers were disenfranchised from taking pride in a finished product and from experiencing the pleasure of a customer well served. This worker alienation led to unions in those days as a way of gaining economic and environmental gains but not necessarily as a means to improve the enterprise. The focus became introspectively adversarial thereby weakening the overall competitive capability of the enterprise to external threats. In recent times, organized labor and management joined forces to defeat a common threat external to the enterprise.

And last, the productivity improvements drove the displacement of hundreds of thousands of jobs thereby disenfranchising workers and sapping the collective intelligence of the enterprise. Performance and dedicated service became irrelevant as economic downsizing resulted from mergers, acquisitions, and new technology. While this phenomenon was once restricted to the seasonal work of the factory, now all workers were equally affected. Professionals and other "white-collar" workers however, do not produce material, but intellectual value. It is difficult to measure the loss of continuity and knowledge created by this simplistic and short term approach to improving the bottom line. Further, it has become quite common for businesses to engage displaced professionals as independent "consultants" because the knowledge needed was no longer resident in the enterprise. Today, business periodicals are full of articles about how to motivate, retain, and restore the team spirit among the survivors of downsizing.

Various methods to reduce worker alienation and sustain productivity gains were implemented under various titles such as: Employee Involvement, Empowerment, Self-Directed Work Teams, and Quality Councils. The positive outcomes from these efforts increased once the organization was viewed as a system requiring cross-functional problem solving and internal alignment of direction.[3] This is where many organizations are today. A cross-functional approach coupled with quality principles certainly leverages organizational impact. However, this thinking may prove still too narrow for today's competetive environment.

The Value Proposition Re-enters the Scene

This parochial view of the enterprise ignores the impact of whole systems thinking. Just as the worker in the factory was concerned about individual components and not the entire product, the same issue exists at an enterprise level. Any business represents but one component in its Value Stream. In today's world, there is often a whole system of supplier specialists whose individual outputs are combined as ingredients to end user product quality. The members of this system are subcontractors, contract manufacturers, raw material manufacturers, and service providers. There is little recognition that they are part of but a single whole system of value added steps to fill the needs of the end user. They often do not know when they have common customers, and even if they do, interaction as a new kind of cross-functional team is minimal. The opportunities to reduce waste in the overall process is no less phenomenal than the organization wide improvements realized when cross functional teams were first established to identify and solve problems together.

Acceptance of a ValueStream Quality System may well represent a competitive model for the next era of commercial organization. By focusing on a model that redefines the enterprise as the entire ValueStream, there exists the opportunity for a competive advantage all but ignored by most businesses. Each connection in the ValueStream that is waste reduced is multiplicative not additive to the end user. Increasingly clear is the notion, that in the coming era, the entire organization needs to be committed to a singular purpose, aligned in strategy and tactics or the enterprise is suboptimized. The organization that is fully aligned and balanced is in a much stronger position to win in the market place.

By developing alliances where suppliers are integrated seamlessly within a single Value Stream and where the process capability is measured and balanced on the basis of the whole system, waste is driven out and the competitive position is improved. The further upstream waste can be prevented the higher the leverage in reducing total system costs. Assume for example, a five stage manufacturing process starting from raw material purchases, component assembly purchases, manufacturing, distribution and sales. A flaw in the raw material connection that reaches the end user has an exponential cost implication compared to the cost of the original material if correct or corrected at its source. What is the cost of a product recall compared to its original cost? Even if the flaw is discovered at various stages prior to the end user, what is the cost comparison of detection and correction further down the Value Stream?

There are many other forms waste can take. These costs are often not as visible or singular in nature but collectively may prove more detrimental to survival. A now obsolete paradigm permits unnecessary costs to remain in the infrastructure of the relationship. Just a few examples of these include: redundant quality inspections and many other duplicative transactions, inaccurate information borne out of non-integrated data networks, higher cost of material purchases, material handling and storage charges, cash consuming inventory balances, and time delays that sap efficiency and dependability.

In addition, there is a degree of industrial integration that may not have been this visible since the days of the craftsman prior to the industrial revolution. It is increasingly common that businesses can be simultaneously both suppliers and customers of each other as part of the same conglomorate or separately. This creates a unique interdependence on mutual success. With merger and divestiture mania that has been prevalent in the 1980s and still into the 1990s, relationships are constantly changing rendering ValueStream optimization a continuous process.

What is the formula for this model called the **ValueStream Quality System**? It is borne out of the integration of two core competencies: *Process Information Technology and Logistics.*

The ValueStream Quality System is primarily a process view of an extended enterprise including customers, stakeholders, and suppliers, it employs a combination of quality and logistics principles designed to eliminate waste throughout the ValueStream.

There are five key value characteristics necessary to develop the competencies necessary for a successful application of the model. They are: information literacy, process orientation, interdependence driven, measurement fluency, and customer-supplier orientation.

With Macrologistics Quality Principles serving as the integrator, the entire supply stream of added value generates efficiencies which create a distinct advantage over disconnected individual enterprises, as shown in the following value strategies.

Value Strategy #1: Information Literacy

Information literacy requires commonality of language and definitions. This is even more important than the capability of the hardware or software deployed. For example, what does "on time" mean? Date shipped or date received? What is "standard" lead time? What are the quality criteria for each of the items in the Value Stream? How is quality measured? How is this information accessible and used by all members of the Value Stream? Are information systems commensurate with the information requirements of the extended enterprise and does it represent a primary communication tool among its members? For example, Wal-Mart recently began offering its vendor-managed inventory software to suppliers to improve product availability to customers, simplify the distribution system, and to reduce overall costs of replenishment. Wal-Mart will provide the training to assist its vendors in providing greater value to the whole system.[4]

Relevant information must be shared openly and be accessible by members of the stream. One attribute of relevancy is "time-boundedness". To the extent that all members of the stream can access information within the same time parameter enables adjustments to be made by those further downstream. Maximum response time facilitates a level of agility that supports the competitive advantage of the specific value chain.

Value Strategy #2: Process Orientation

Process orientation reflects an understanding of the flow of how the work is accomplished with a critical eye toward improvement. A process is defined as the steps that convert an input into an output. This is different from traditional organizations which have been viewed as a sequence of functional activities. Processes can be mapped to reflect how work and information flow through the enterprise. Process mapping and flowchart techniques used to document a process or system by a visual representation of work and information flow.

Quality tools are used to diagnose unnecessary, redundant, or ineffective transactions within the process. In addition, they serve as a means to identify, plan, implement, and modify process improvements. The ability to smoothly integrate flexibility to change is key in an environment where adaptation to change means survival and success.

Value Strategy #3: Interdependence Driven

Interdependence driven refers to a whole systems perspective and a recognition that the entire Value Stream is affected by the actions of each of its members. Given the criticality of these relationships, selection of suppliers and channels to serve end users of products becomes a critical part of developing competitive advantage. This is particularly true in today's world when outsourcing is a common component of the Value Stream. The notion of a virtual factory is a reality on a global basis. Where there exists common goals albeit for different purposes, a basis for strategic alliances may be possible. When strategic alliances are created, focus is enhanced by the concentration of each member on a portion of the process. The vendor selection process requires less ongoing activity and the procurement cycle may be less volatile. Access to capital may be diversified. Further, those who excel in needed processes may very well be the candidates selected to apply their superior competency within the extended enterprise. This combination possesses significant strength toward achieving a system of excellence that has the makings of competitive advantage.

All too frequently, however, the terms "partner" and "strategic alliance" are abused and represent nothing more than sales hype. Learning how to recruit and evaluate suppliers who have demonstrated clear and compatible goals is necessary as well as extremely difficult.

Trust can only emerge when all those organizations who participate in the entrerprise can fulfill their individual goals as a result of the alliance. Last, this trust is a precondition to the access and sharing of critical and confidential information necessary for success.

Value Strategy #4: Measurement Fluency

Measurement fluency involves the creation of clear and objective measures that must be linked to strategy. The linkage of goals to measures is the vehicle to establish clear ownership.and to evaluate progress. Outcome, Process, and Predictive Measures provide a basis for common language to be integrated with the information system as well. Outcome measures are historical in nature. They reflect results to date. Process are Just-in-Time measures which monitor the process as it occurs. Predictive measures are useful in terms of cause-and-effect relationships within the Value Stream. This line of sight from results backward to Just-in-Time and ultimately predictors of performance provides the opportunity to anticipate and proact. An understanding of the relationships between measures and the ability to use them as tools to achieving results is key. The measures also reflect where attention needs to be placed and in a timely fashion. As such, the measures serve as a critical link in enabling the members of the Value Stream to communicate effectively. Clarity of communication along with speed represent important components of competitive advantage.

Value Strategy #5: Customer–Supplier/Stakeholder Orientation

Customer–supplier/stakeholder orientation represents a view that the customer is more than just the next step in the value-added process. In addition, it implies that the customers within the ValueStream are responsible to provide timely feedback about expectations, specifications and performance. In addition there is a mutual accountability to improve the effectiveness of the supplier-customer relationship within the Value Stream. This perspective also includes a focus on how all the members of the Value Stream can improve the value added to the end-user customer. Finally, these stakeholder groups must also be considered when analyzing business processes: owners, investors, employees, governments and community groups.

In the future, each enterprise will be required to select what it's core competencies are and choose how they fit into the ValueStream leading to the end user. Historically many companies vertically integrated to control cost and quality but in fact this accomplished just the opposite. These conglomerates became too complex, required too many resources to be best at everything, and had costs that increased, sometimes undetected, until substantial waste occurred.

Figure 8.1 is a matrix reflecting the contrasting perspectives which have evolved toward a whole-systems approach to business. What began as primarily a product-driven introspective view of quality focused on cost reduction, has been extended toward a full performance view of a single enterprise. Initially, quality efforts were focused on production, a single component of the organization. Root cause analysis and the use of data-driven problem solving revealed that production quality issued were sometimes caused by design issues or others outside of the production process.

The globalization of manufacturing and the growth of service industries as the growth sector has further extended the view of quality to incorporate this now dominant component of our economy.

Value Strategy #6: Quality Council

Further, we have come to understand that quality is a function of the entire process and not simply a statistical problem solving device for technical inconsistency.

Customers and stakeholders are now understood to mean both internal and external creating a new level of accountability among members within an organization to each other. This is an important step to truly understanding the organization as a system, requiring full integration and balance. Recently, some organizations have learned that by incorporating vendor systems into the process design, additional quality enhancements can be achieved, of course not all stakeholders participate in every process.

The complexity to create balance throughout the extended enterprise requires a highly developed understanding of process management. One mechanism to accomplish this task is the creation of a Quality Council. This council exists as a steering team as well as a forum to establish direction, priorities, resources, and to discuss ways to assure that the quality goals of the enterprise are met. Also, understanding which stakeholders needs must be met and having this information easily helps avoid confusion and improves focus.[14]

Because several organizations may be involved, it is important to consider the variety of directions and priorities of these independent enterprises.

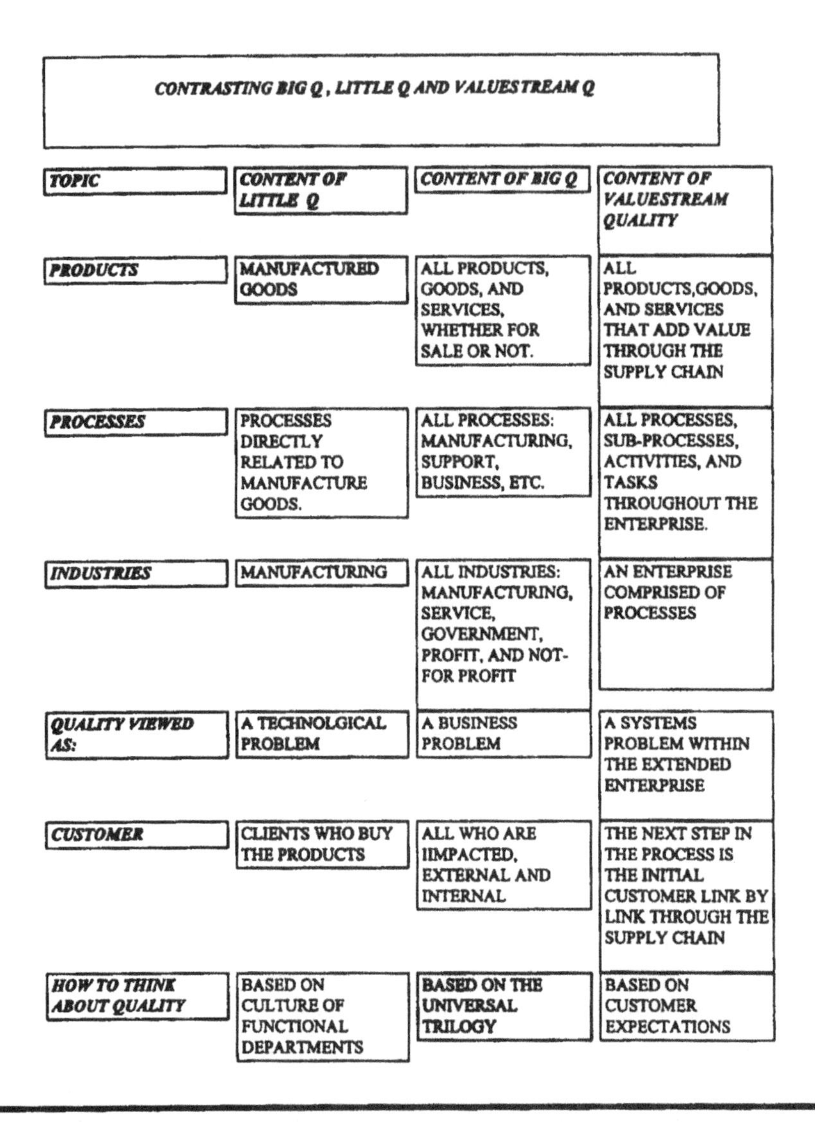

CONTRASTING BIG Q, LITTLE Q AND VALUESTREAM Q			
TOPIC	CONTENT OF LITTLE Q	CONTENT OF BIG Q	CONTENT OF VALUESTREAM QUALITY
PRODUCTS	MANUFACTURED GOODS	ALL PRODUCTS, GOODS, AND SERVICES, WHETHER FOR SALE OR NOT.	ALL PRODUCTS,GOODS, AND SERVICES THAT ADD VALUE THROUGH THE SUPPLY CHAIN
PROCESSES	PROCESSES DIRECTLY RELATED TO MANUFACTURE GOODS.	ALL PROCESSES: MANUFACTURING, SUPPORT, BUSINESS, ETC.	ALL PROCESSES, SUB-PROCESSES, ACTIVITIES, AND TASKS THROUGHOUT THE ENTERPRISE.
INDUSTRIES	MANUFACTURING	ALL INDUSTRIES: MANUFACTURING, SERVICE, GOVERNMENT, PROFIT, AND NOT-FOR PROFIT	AN ENTERPRISE COMPRISED OF PROCESSES
QUALITY VIEWED AS:	A TECHNOLGICAL PROBLEM	A BUSINESS PROBLEM	A SYSTEMS PROBLEM WITHIN THE EXTENDED ENTERPRISE
CUSTOMER	CLIENTS WHO BUY THE PRODUCTS	ALL WHO ARE IIMPACTED, EXTERNAL AND INTERNAL	THE NEXT STEP IN THE PROCESS IS THE INITIAL CUSTOMER LINK BY LINK THROUGH THE SUPPLY CHAIN
HOW TO THINK ABOUT QUALITY	BASED ON CULTURE OF FUNCTIONAL DEPARTMENTS	BASED ON THE UNIVERSAL TRILOGY	BASED ON CUSTOMER EXPECTATIONS

Figure 8.1 Big Q, Little Q, and ValueStream Q

both compatible and synergistic if the Council concept is to succeed. An additional complication is that system optimization does not guarantee the optimization of each component within the system. One should not expect an independent vendor to willingly suboptimize their organization without some sharing in the improvement of the entire chain. This reciprocity must be understood and built into the infrastructure early if this model is to be successful.

TOPIC	CONTENT OF LITTLE Q	CONTENT OF BIG Q	CONTENT OF VALUESTREAM QUALITY
QUALITY GOALS ARE INCLUDED:	AMONG FACTORY GOALS	IN COMPANY BUSINESS PLAN	IN AN ALIGNED BUSINESS PLAN THROUGHOUT THE SUPPLY CHAIN TO THE END USER.
COST OF POOR QUALITY	COSTS ASSOCIATED WITH DEFICIENT MANUFACTURED GOODS.	ALL COSTS THAT WOULD DISAPPEAR IF EVERYTHING WERE PERFECT.	ALL COST THAT WOULD DISAPPEAR IF THE SYSTEM WERE BALANCED AT ITS PROCESS CAPABILITY.
IMPROVEMENT IS DIRECTED AT:	DEPARTMENTAL PERFORMANCE.	COMPANY PERFORMANCE.	WHOLE SYSTEMS BALANCE OF THE EXTENDED ENTERPRISE.
EVALUATION OF QUALITY IS BASED MAINLY ON:	CONFORMANCE TO FACTORY SPECIFICATIONS, PROCEDURES.	RESPONSIVENESS TO CUSTOMER NEEDS.	SYSTEM OPTIMIZATION.
TRAINING IN MANAGING QUALITY IS:	CONCENTRATED IN THE QUALITY DEPARTMENT.	COMPANY WIDE.	THROUGHOUT THE EXTENDED ENTERPRISE.
COORDINATION IS BY:	THE QUALITY MANAGER.	A QUALITY COUNCIL OF UPPER MANAGERS.	A QUALITY COUNCIL WHOSE MEMBERS ARE REPRESENTATIVES OF THE ENTIRE VALUE STREAM.

Figure 8.1 Process Mapping (continued)

Value Strategy #7: Three Dimensions of Quality

One additional purpose of the Council is to economically distribute training and development resources across a broader base. Not only is this more cost effective, it also reinforces a common language, skill development and improved communcation within the ValueStream.

Today, as never before, we must develop speed and agility in the acquisition and application of information throughout the ValueStream in order to survive in an environment where competition can emerge literally from any place on the planet.

Conclusion

Managers are now beginning to realize that balancing the needs of all the stakeholders — customers, suppliers, owners, employees, governments and community groups — must be balanced. Understanding each groups needs and balancing them, without jeopardizing the well-being of the others is an important determinant of whether an activity adds value to the ValueStream.

Since stockholders seldom come into direct contact with each other, the ValueStream Quality Management model can play a pivotal role in developing the process required to balance their needs.[6]

References and Endnotes

1. Tischler, William, *Understanding and Applying Value-Added Assessment*, ASQC Quality Press, 1966, pp. 5–7.
2. Voehl, Frank, *History of Total Quality*, St. Lucie Press, Boca Raton, FL, 1995.
3. DeRose, Louis, *The Value Network: Integrating the Five Critical Processes*, New York, Amazon, 1994.
4. Hammer, Michael and Steven Straton, *The Reengineering Revolution: A Handbook*, New York, Harper Business, 1995.
5. For further information on the role of Quality Controls, see *Quality Control Handbook* by Strategy Associates, Coral Springs, FL, 1996, pp. 15–25.
6. Felkins, Patricia, B. J. Chakris, and Kenneth N. Chakris, *Change Management: A Model for Effective Organizational Performance*, New York, Quality Resources, 1996. Voehl & Vengrow, "Value as the New Quality Wave", Quality Observer, 1998.

Whirlpool Leverages its Resources to Become a World-Class Competitor 8.1

When David Whitwam became CEO of Whirlpool in 1987, the firm was mired in a war of attrition in the U.S. market, a war in which cost and quality were the only weapons and declining margins the only prize. Whitwam recognized, however, that the firm had to be a global competitor and that this would require leveraging global capabilities and eliminating regional fiefdoms and inadequate ways of satisfying customers.

In 1989 in a daring move, he purchased N.V. Philips's floundering European appliance business, a move that catapulted Whirlpool into the number one position in the worldwide appliance industry. Then he chose to transform the two firms into a unified, customer-driven organization that could use its combined talents to create chaos in the global marketplace. As a result, Whirlpool has set the standard for the industry in terms of new-product innovation ad industry price structure.

Whitwam asserts, "The only way to gain lasting competitive advantage is to leverage your capabilities around the world so that the company as a whole is greater than the sum of its parts. Being an international company — selling globally, having global brands or operations in different countries isn't enough.To me, 'competitive advantage' means having the best technologies and processes for designing, manufacturing, selling and servicing your products at the lowest possible costs. Our vision of Whirlpool is to integrate our geographical businesses wherever possible.We want to be able to take the best capabilities we have and leverage them in all of our operations worldwide."

To achieve these goals, he adds, "You must create an organization whose people are adept at exchanging ideas, processes, and systems across borders, people who are absolutely free of the 'not-invented-here' syndrome, people who are constantly working together to identify the best global opportunities and the biggest global problems facing the organization."

"Our strategy is based on the premise that World-Class cost and quality are merely the ante — the price of being in the game at all. We have to provide a compelling reason other than price for consumers to buy Whirlpool-built products. We can do that only be understanding the consumer better than anyone else does and then translating our understanding into clearly superior product designs, features, and after-sales support." His goal is for consumers to prefer the Whirlpool brand because it offers greater overall value than competing products. He believes that achieving that goal requires taking a giant step back from the business and rethinking who their customers are and what their needs are. He suggests that to many this may not sound earth-shattering, but it is because it means rethinking the very nature of the business.

He continues, "all of us in this industry have been telling ourselves that we're in 'the refrigerator business,' 'the washing-machine business,' or "the range business.' None of us saw a great deal of room for product innovation, which is undoubtedly why there hasn't been radical innovation in thirty years, apart from the microwave oven and the trash compactor. If you want to open the door to imagination and innovation, isn't it more useful to think of 'the fabric-care business,' and 'the food-preservation business'?"

He believes that the starting point isn't the existing product but the function consumers buy products to accomplish. "When you return to first principles, the design issues dramatically change. The microwave couldn't have been invented by someone who assumed he or she was in the business of designing a range. Such a design breakthrough required seeing that the opportunity is 'easier, quicker food preparation,' not 'a better range.'"

Whirlpool is applying these broader definitions in very direct ways. For example, organizationally, they have created what they call an advanced product-development capability to serve markets around the world. Its mission is to look beyond traditional product definitions to the consumer processes for which products of the future will have to provide clear benefits.

He adds, "take 'the fabric-care business,' which we used to call the 'washing-machine business.' We're now studying consumer behavior from the time people take off their dirty clothes at night until they've been cleaned and ironed and hung in the closet. What are we looking for? The worst part of the process is not the washing and drying. The hard part is when you take your clothes out of the dryer and you have to do something with them — iron, fold, hang them up. Whoever comes up with a product to make this part of the process easier, simpler, or quicker is going to create an incredible market."

 Source: Regina Fazio Maruca, "The Right Way to Go Global: An Interview with Whirlpool CEO David Whitwam," *Harvard Business Review* (March–April 1994), pp. 134–145.

Hewlett-Packard's Kanban Approach to Employee Education 8.2

An important trend in knowledge management is analogous to the Kanban approach to inventory management. The Kanban approach is named after the Kanban cards used in Toyota's production lines. In this system, when a team of workers needs a specific part it sends a Kanban card to internal suppliers requesting that part. Because it sends the card only when it needs parts, supply is based directly on demand and there is no need to hold more inventory than is needed at a particular time. A similar shift to just-in-time delivery of knowledge to employees is occurring in several companies. The basic idea is to do away with classrooms and put education on line, on demand. Hewlett-Packard is leading the way to improved educational services at a small percentage of previous costs.

As discussed in *Innovate or Evaporate 4.1,* at HP innovation is a corporate strategy. In human resources, for example, the firm has adopted a Kanban approach to training. "We're constantly pushing to blur the lines between learning and doing the job," says Susan Burnett, manager of worldwide sales force development.

A few years ago HP found that virtually all of its sales training was done in the classroom. Sales reps were spending three weeks a year in classrooms rather than with customers. Moreover, of the time spent with customers, about three-fourths was spent transferring catalogs and reports from the central office to the customers' offices. Compounding problems for the sales force was a shrinking product life cycle from eighteen months to as little as six months. At the same time, the technological complexity, specs, and applications of products were growing by leaps and bounds. Finally, administrative overhead was being cut just when all this information was approaching overload levels.

The solution was the Hewlett-Packard Interactive Network (HPIN) that Tom Wilkins, R&D manager for media technologies, helped create. The network has been so cost-effective that it has become a profit center. Before 1990, whenever HP rolled out a major new product, it would bring its 950 sales reps to a conference center for a day or two of training, at a cost of about $5 million. The use of HPIN has reduced that cost to $80,000, a reduction of more than 98%, with no loss in effectiveness.

Sources: Lewis J. Perelman, "Kanban to Kanbrain," *Forbes ASAP* (June 6, 1994), pp. 85–95.

Return on Quality 8.3

The premise of Return on Quality (ROQ) is that very few companies can accurately translate subjective goals relating to quality and customer satisfaction into hard numbers which can be linked to specific logistics improvement programs. The book entitled *Return on Quality,* describes a methodology to measure corporate quality efforts and to quantify "their return on investment." It's authors stress that its undeniable that organizations desperately need such a system; the question is how to provide one that works. The ability to accurately measure results provides the basis for accurate forecasting as well as providing for the logistical deployment of resources. The key problem, they point out, is that placing a value on feelings such as satisfaction proves to be a difficult assignment under even the best conditions.

The authors attempt to solve this dilemma by first analyzing the decision paths where quality leads to profits. These paths, which can include both product performance as well as both process and service features, leads to customer satisfaction, which leads to increased market share, which leads to improved profitability and competitive advantage. The book contains surveys, user studies, quality tools, instructions for conducting focus groups, as well as numerous other techniques for data gathering at each step of the process. They also show how to use statistical analysis to translate findings into hard numbers, which are then applied to the model to rationalize decision making as well as helping in the strategic quality planning process. The promise of the book is the creation of the "high-performance business."

A unique addition to the book is an added value return-on-quality software system which computerizes the decision making and logistics planning process. The reader can use a demonstration disk, available from the publisher for an additional fee, to process the statistics gathered from customers into the proper screens, from which the software generates models. According to reviewer Ted Kinni, about 25% of the book describes this software and several case studies illustrate this illusive concept of "return on quality" in action.

This book is an impressive work, although it has some weaknesses. The concept of translating customer values into numbers is a difficult task "fraught with peril and the values generated by the model are only as good as its assumptions and data fed into it." In bold print they issue this warning: "Remember this warning: decision support systems supply input to a manager's decision. They don't make the decision." In other words, caveat emptor — let the buyer-reader beware.

Source: Rust, Roland, Zahorik, *Anthony and Keiningham,* Timothy Probus Publishing, New York, 1994.

Texaco Redefines the Gasoline Retail Industry 8.4

Texaco wasn't the first gasoline retailer to combine filling stations with convenience stores, but in 1991 it became the first of the major oil firms to do so when it introduced its Star Marts. Texaco has also been upgrading its refineries and creating both new and improved products, such as cleaner-burning fuels. The results have been nothing less than sensational. From last among the big six oil firms in 1987, by 1993, Texaco had moved into fourth place.

In the late-1980s, following its filing for bankruptcy protection and an unsuccessful run at the company by corporate raider Carl C. Icahn, former CEO James W. Kinnear began a major strategic overhaul of the firm. He slashed the payroll by almost a third, cut other costs, and sold off unprofitable businesses. He boosted Texaco's oil exploration efforts, sought to improve its refining operations, and focused on making the firm's marketing more effective. The latter required both "big bang" and continuous innovation.

In a stagnant, highly competitive market, where a difference of 1¢ per gallon will cause customers to shift companies, Texaco has steadily been building customer loyalty. The Star Mart stores are a major reason. Other oil firms, seeing Texaco's success, have begun to imitate this concept. But Texaco is ready for them. It is increasing the size of many of its 6000 stores and adding fast-food franchises such as McDonald's and Dunkin' Donuts to many of them. It is also testing the addition of quick-lube facilities at is stations. "All we want to do is to get you into our stations," notes Donald H. Schmude, president of Texaco Refining & Marketing. To further entice motorists, Texaco offers rebates on gasoline purchases.

The company is also moving ahead with its refining improvements, planning to invest another $240 million to make its upstream operations nearly self-sufficient in the production of oxygenates, which help limit auto pollution emissions. This adds to the over $2.5 billion in improvements in refinery technology that Texaco and its downstream partner, Saudi Arabian Oil Company, have made since 1988. Texaco is also pursuing an innovative strategy of swapping filling stations with other oil companies in order to get as many stations as possible close to its own seven refineries. (In this industry, where 1 cent a gallon is critical, reduced transportation costs make a huge difference.) Recently, for example, Texaco swapped fourteen outlets in St. Louis for fifteen Amoco stations in the Rocky Mountain region. Its cleaner System-3 fuels are being introduced in international markets, and other innovations are likely to follow.

© James M. Higgins. *Source:* Tim Smart. "Pumping Up at Texaco," *Business Week* (June 7, 1993), pp. 112–113.

Keiretsu in America 8.5

The author of *Keiretsu in America* is a well-known business writer who presents a concise and intriguing look at one of the key features of Japan's success — an integrated Marketing & Sales/Purchasing/Supplier Logistics Management system: *keiretsu*. The opening frame sets the stage with a simple well-directed statement: "Corporate Communism? Industrial war machines? As U.S. business comes to terms with dealing with Japan Inc. on home turf, it's time we understood how our new neighbors do business." The reason: according to Kinni, Japanese business interests hold a sizable stake in over 1500 U.S. factories, a fact which is confirmed by the Japan External Trade Organization.

Keiretsu, is the Japanese system of conglomerates which cross-market and trade heavily with one another, dating back to the Meiji Restoration of 1868. The author quotes Robert Kearns in describing how an American *keiretsu* structured along the lines of the Mitsubishi or Sumitomo group might operate: "Such a group would be worth close to a trillion dollars. Each of the 30 or so lead companies would own a piece of each other, would do business among themselves and meet once a month for lunch and discuss matters."

Keiretsu members share in the economies of large-scale operation, says the author, such as low-cost capital at rates of 0.5 to 1.5% interest. Membership in a *keiretsu* virtually guarantees a market for one's goods, since other member companies own large stakes in each other's companies, and the value of their investments depends on the long-term success and growth of the member firms. According to Kinni, "the *keiretsu* system is an ideal structure for rapid and secure economic growth and a major reason for Japan's economic success since World War II." After describing how the system operates, the author turns to examples of *keiretsu* way, which means allowing a foreign group to eventually own about 30% of one's company stock, and to some extent, dictate the organization's future. Some U.S. companies, like Timken Co., are building structures reminiscent of the *keiretsu* on their own, such as the "supplier city" in Perry Townhip, Ohio, which will bring its suppliers within arm's reach. What does the future hold?

Invasion or evolution? The author uses observers and "experts" to sum up the final arguments. "As the Japanese investments in this country mature, their plants and equipment will age, their employees will grow more expensive, and the playing field will level. Perhaps the *keiretsu* will learn a new respect for the individual and will begin to temper its authoritarian structures." On the other hand, he believes that U.S. corporations can learn much from the business practices of the *keiretsu,* with its efficient cross-marketing and supplier relationships.

The best way to come to terms with *keiretsu,* Kinni concludes, is to think of it as an immigrant, not an invader, with on one hand gifts to offer and lessons to learn — for those wise and gracious enough to know how to use them. As America moves further along in its understanding of the "extended enterprise", businesses will see themselves as members of a global network, whose aim will be to optimize the value chain through mutual relationships built on trust.

Source: Theodore B. Kinni, *Quality Digest,* December 1992, pp. 24–31.

9 Benchmarking Using Critical Success Factors to Implement Macrologistics Strategies

The concept of Critical Success Factors (CSFs) was first popularized by Jack Rockart of Massachusetts Institute of Technology.[1] CSFs for any business are the limited number of areas in which satisfactory results will ensure successful competitive advantage; in other words, the areas where things must go right if the effectiveness of the organization is to flourish and greater market share is to be achieved.

Organizations need to define a breakthrough (stretch) CSF that hasn't yet been attained by the competition. For example, Federal Express used the CSF of delivery time to achieve competitive advantage. Their strategy was to exceed customer expectations at a reasonable cost with exceptional service quality. This formula allowed them to expand market share and their CSF approach became an example of a Macrologistics strategy.

> "When XYZ is the best of the best, there's plenty to learn and apply. When Southwest Airlines, the best of the best in the airlines business, wanted to improve the cleaning and refueling of its planes and turn them around quickly, it benchmarked its performance against the Indy 500 pit crews."
>
> Keen and Knapp

Critical success factors can be categorized as either "monitoring," "building" or "benchmarking" types. The more competitive pressure for current performance that the chief executive feels, the more his or her CSFs tend toward monitoring current results. The more that the organization is insulated from economic pressures or decentralized, the more CSFs become oriented toward building for the future through major change programs aimed at adapting the organization to a perceived new environment. The more the organization has identified the need for comparison and adaptation of other organizations' best practices, the more CSFs become oriented towards Benchmarking the operations of others.

Criteria for CSFs

In general, at least five criteria can be used on the corporate level to determine which factors are critical to the effectiveness of an organization:

1. Overall impact on performance measures, such as profitability, cash flow, return on investment and competitive positioning.
2. Overall relationship to the strategic direction and issues, such as differentiation, turnaround and segmentation.
3. Relationship to more than one business activity.
4. Relationship to stages in product or organizational life cycle, such as introduction, growth and decline.
5. Overall impact involving large amounts of capital and resources in relation to activities of the organization.

Once the corporate level CSFs are established, each department is encouraged to identify and align with indicators that can be used to measure its contribution. In this way, a corporate-wide system of indices can be linked and tied into the organization's performance measurement system.

Performing the CSF Interviews

The MIT research team's experience in conjunction with the Florida Power and Light study of 1980–1981 found the critical success factors (CSF) approach is highly effective in helping executives to define their significant information needs.[2] Equally important, it has proved efficient in terms of the interview time needed (from three to six hours) to explain the method and to focus attention on information needs. Most important, executive response to this method has been excellent in terms of both the process and its alignment to outcome.

The actual CSF interviews are usually conducted in two or three separate sessions. In the first, the executive's goals are initially recorded and the CSFs that underlie the goals are discussed. The interrelationships of the CSFs and the goals are then talked about for further clarification and for determination of which recorded CSFs should be combined, eliminated or restated. An initial cut at measures is also taken in this first interview.

The second session is used to review the results of the first, after the analyst has had a chance to think about them and to suggest "sharpening-up" some factors. In addition, measures and possible reports are discussed in depth. Sometimes, a third session may be necessary to obtain final agreement on the CSF measures-and-reporting sequence.[3]

CSFs and Competitive Marketing Analysis

Overall, CSFs differ from company to company and from manager to manager and go beyond the norm of control to include the idea of improvement and breakthrough as well as competitive strategy.

"Next to knowing what your customers want, the most important thing is to know what your competitors are doing." This remark by John Rhode, Vice President of Marketing and Planning with Combustion Engineering's Industrial Group, summarizes the case for competitive intelligence. Michael Porter of the Harvard Business School, one of the key players in his field, states that competitive strategy involves positioning a business to maximize the value of the capabilities that distinguish it from its competitors. It follows that a central aspect of strategy formation is a competitor analysis that strives to use every iota of data gathered. He argues that one of the key functions of the corporate measurement system is to provide information profiles on each key competitor. If too much work is done on data *pluming* alone, the organization develops the habit of not acting on data, which causes piles of unused data to get developed.[4]

Competitor analysis has been criticized by some practitioners who think that his discussion of competitor analysis as discussed isn't very realistic and it would be extremely difficult to collect all of the information he considers essential. Another view is that there's been far too much emphasis on strategy and rivalry and not enough on operating details and doing things well. In a *Harvard Business Review* article titled "Hustle as Strategy," Amar Bhide writes:

> "Opportunities to gain lasting advantage through blockbuster strategic moves are rare in any business. What mostly counts are vigor and nimbleness. These traits are always needed and always important, yet strategic planning theologians largely ignore them. Countless companies in all industries, young or old, mature or booming, are finally learning the limits of strategy and concentrating on tactics and execution. In a world where there are no secrets, where innovations are quickly imitated or become obsolete, the theory of competitive advantage may have had its day. Realistically, ask yourself, if all your competitors gave their strategic plans to each other, would it really make a difference?

Collecting various kinds of data that may be relevant — anything from trade rumors and financial statistics to product specifications and news of plant construction — and then selecting, interpreting, and presenting the data as information to be used in decision making is vital. The benchmarking consulting firm of Michael M. Kaiser provides this observation:

> "The most common reason why firms have not performed a more Competitive analysis is the perceived difficulty of finding the required data. Competitive Analysis are treasure hunts in that everyone is trying to uncover enough data to answer questions which relate to another firm's resources and personality. While it is impossible to be sure that the conclusions are correct, collecting as much pertinent information as possible increases the odds of accuracy (but one has to guard against spending too much time collecting data). A superior competitive analysis is the one with the most insights, not the one with the most facts. The goal is to gather the data which supports the generation of these insights."

For Macrologistics practitioners, the key to using this technique of customer analysis is in effectively developing breakthrough execution tactics that differentiate the organization on cost or quality factors. Otherwise, we concur that competitor analysis by itself is not enough.

Sources of CSFs

CSFs are applicable to any company or enterprise operating in a particular *industry.* Yet many experts have emphasized that a management control and measurement system also must be tailored to a particular *company's value chain.* This suggests that there are other sources of CSFs than the industry alone. The MIT team, headed by Jack Rockart, identified four prime sources of logistics-based critical success factors:

1. **Structure of the particular industry.** Each industry by its very nature has a set of CSFs that are determined by the characteristics of the industry itself. Each company in the industry must pay attention to these factors.
2. **Competitive strategy, industry position and geographic location.** Each company in an industry is in an individual logistical situation determined by its history and current competitive strategy. For smaller organizations within an industry dominated by one or two large companies, the action of the major companies will often produce new and significant problems for the smaller companies. The competitive strategy for the latter may mean establishing a new market niche, getting out of a product line completely, or merely redistributing resources among various product lines. Thus for small companies a competitor's strategy is often a CSF. Just as differences in industry position can dictate CSFs, differences in geographic location and in strategies can lead to differing CSFs from one company to another in an industry.
3. **Environmental factors.** As the gross national product and the economy fluctuate, as political factors change, and as the population waxes and wanes, critical success factors can also change for various institutions. At the beginning of 1973, virtually no chief executive in the U.S. would have listed "energy supply availability" as a critical success factor. Following the oil embargo, however, for a considerable period of time this factor was monitored closely by many executives — since adequate energy was problematical and vital to organizational bottom-line logistical performance.
4. **Temporal factors.** Internal organizational considerations often lead to temporal critical success factors. These are areas of activity that are significant for the success of any organization for a particular period of time because they are below the threshold of acceptability at that time. As an example: for any organization, the loss of a major group of executives in a plane crash would make the "rebuilding of the executive group" a critical success factor for the organization for the period of time until this was accomplished.[3]

Items one to three covered above suggest a strong **alignment** as well as **integration** role for Benchmarking in the competitive intelligence and strategic planning process. In its classical form, it is defined as modeling and monitoring of the best practice regardless of industry.

Once the corporate level CSFs are established, each department is encouraged to identify indicators that can be used to measure its contribution. In this way, a corporate-wide system of indices can be linked and integrated into the organization's performance measurement system. However, a process is needed to make this happen and that is what adaptive engineering does very well — to follow through on CSFs on a continuous improvement basis. A Macrologistics CSF transforms the service and product value of the organization in the marketplace in which it competes, thereby helping to increase the market share potential.

The AT&T Story: Alignment Gone Astray

The relationship between CSFs, competitive positioning and Policy Deployment is best illustrated by AT&T's abortive foray into the computer industry. In 1983, AT&T believed it was poised to power its way into the computer industry. The company bragged about the fact that its Bell Laboratories research and development facility was the largest in the world, and that in supplying equipment for its telephone network it had as much experience manufacturing computers as did IBM. It had everything it needed — it could design computers and it could make them — and was ready for a head-on confrontation with IBM for computer industry supremacy.

Five years later, that scenario seemed ludicrous. AT&T's attempts to enter the computer industry — much less dominate it — had failed miserably. Year by year, the company lowered its objective: from confronting IBM, to beating DEC and others for the number two position, to securing a significant niche (way short of being number two) to somehow, some way, getting a foothold in some corner of the market. As the company regrouped yet again in 1988, one thing was clear: no future success would make up for the years of effort and the billions of dollars lost in the meantime.

What went wrong? How did the rosy AT&T projections lead to such disaster? The answer lies not so much in what AT&T did, as in what it *didn't* do: align its CSFs.

First, AT&T failed to assess what it really took to win in the computer business. A strategic examination of the industry environment and the various competitors — the scan portion of Policy Deployment — would have revealed that sales, distribution, and service were the critical success factors, not R&D and manufacturing. By the mid-1980s almost anyone could make a computer. The key was having a large, high-quality, experienced sales and service force to control the corporate customer and convince him to buy your products. IBM understood this concept and adapted its policies thoroughly, but AT&T did not and targeted outmoded goals that were already superseded in the computer industry. Had AT&T performed Strategic Benchmarking and combined this with a detailed competitor analysis, the results might have been different.

Second, AT&T failed to rigorously and realistically measure where it stood vs. the competition relative to the critical logistical factors, a process called *operational benchmarking*. Such an analysis would have revealed several big differences between the sales and service functions of IBM (or DEC) and those of AT&T:

1. Whereas AT&T had very large sales and logistics service forces, because of AT&T's historical monopoly, it had relatively little experience in competitive markets.
2. Much more importantly, AT&T's sales force sold telecommunications products and services but knew nothing about how to sell computers! Similarly, the experience of AT&T's service force in maintaining telecommunications systems had almost no applicability to the type of logistical service relationships common in the computer industry.
3. A good operational benchmarking analysis also would have demonstrated that AT&T's overall costs were way out of line, meaning that even if AT&T had been successful in capturing market share and revenue in the computer industry, it was unlikely that AT&T would have earned a satisfactory profit doing so.

The other missing ingredient in this situation was the inability to correctly understand the customer needs and wants, and to listen to the voice of the customer. In other words, they should have had a better system for surfacing and aligning their CSFs. Such an approach is found with Quality Function Deployment (QFD) which can be effectively used to surface CSFs and develop Breakthrough strategy while creating policy alignment, as described in Chapter 3.

The actual CSF interviews are usually conducted in two or three separate sessions. In the first, the executive's goals are initially recorded and the CSFs that underlie the goals are discussed. The interrelationships of the CSFs and the goals are then talked about for further clarification and for determination of which recorded CSFs should be combined, eliminated or restated. An initial cut at measures is also taken in this first interview.

The second session is used to review the results of the first, after the analyst has had a chance to think about them and to suggest "sharpening-up" some factors. In addition, measures and possible reports are discussed in depth. Sometimes, a third session may be necessary to obtain final agreement on the CSF measures-and-reporting sequence.

From Control Systems to Strategic Systems

In an attempt to overcome some of the shortcomings of the major approaches involved in developing relevant logistics-based information, the CSF method focuses on *individual managers* and on each manager's *current information needs* — both hard and soft. It provides for identifying managerial information needs in a clear and meaningful way. As CSFs are developed at various levels, new problems surface and are brought to light. Moreover, the process takes into consideration the fact that information needs will vary from manager to manager and that these needs will change with time for any particular manager. Furthermore, many corporate level CSFs don't translate into divisional CSFs very well and a standard approach is needed to sorting this out.

This standard approach is based on the concept of the "success factors" first discussed in the management literature in 1961 by D. Ronald Daniel of McKinsey & Company. Daniel, in introducing the concept, cited three examples of major corporations whose logistics information systems produced an extensive amount of information. Very little of the information, however, appeared useful in assisting managers to better manage the value chain.

Once the CEO knows that the organization must change, the question becomes how does he/she create the change to happen. In many cases, the CEO can't get the change to happen and all they are stuck with is an expensive information system. Rockart assumes that to identify the CSF is enough and the CEO becomes trapped by the bureaucracy and resistance to change.

To draw attention to the type of information actually needed to support managerial activities, Daniel turned to the concept of critical success factors. The following are some examples from several major industries that he identified and integrated into a standard process: In the automobile industry, styling, an efficient dealer organization, and tight control of manufacturing cost are paramount; in food processing, new product development, good distribution and effective advertising are the major success factors; in life insurance, the development of agency management personnel, effective control of clerical personnel and innovation in creating new types of policies spell the difference."

As Daniel notes, critical success factors support the attainment of organizational goals or objectives. Goals represent the end points that an organization hopes to reach. Critical success factors, however, are the areas in which good performance is necessary to ensure attainment of those goals. It is this integration with corporate goals and objectives that gives value and meaning to the strategic planning process, when coupled with a Macrologistics Management strategy. Without focusing on the time/place economies of the processes, companies often overlook this area of achieving greater competitive advantage without major cost. Using Macrologistics helps find and capitalize on value-added opportunities.

Daniel focused on those critical success factors that are relevant for any company in a particular industry, and he used the supermarket logistics model as an illustration. Supermarkets have four industry-based CSFs. These are having the right product mix available in each local store, having it on the shelves, having it advertised effectively to pull shoppers into the store, and having it priced correctly — since profit margins are low in this industry. Supermarkets must pay attention to many other things, but these four areas are the underpinnings of successful operation and, combined with JIT approaches, can lead to competitive advantage.

Writing a decade later, Anthony expanded this concept into the work on the design of management control systems. They emphasized three "musts" of any such system: "The control system *must* be tailored to the specific industry in which the company operates and to the specific strategies that it has adopted; it *must* identify the 'critical success factors' that should receive careful and continuous management attention if the company is to be successful; and it *must* highlight performance with respect to these key variables in reports to all levels of management."

One of the barriers to implementing Macrologistics CSFs is that the Logistics Department doesn't have the relevant reports and doesn't spend much time with senior management. This is also true of the Purchasing function as well. They are looked at as "plumbing" only, and get attention when something breaks.

Tailoring CSFs to a Company and Its Competition

While continuing to recognize industry-based CSFs, Anthony et al. thus went a step further. They placed additional emphasis on the need to tailor management planning and control systems to both a company's particular strategic objectives and its particular managers. Thus, performance measurement systems must report on those success factors that are perceived by the managers as appropriate to a particular job in a particular company. In short, CSFs differ from company to company and from manager to manager and go beyond the norm of control to include the idea of improvement and breakthrough as well as competitive strategy.

Why Benchmarking Is So Important

Historically, Western organizations have tended to look inward rather than outward, and to be secretive about their operations. Benchmarking, on the other hand, is the calibration of the organization's performance with that of another organization representing the "best in the field." It takes firms away from the "not invented here" syndrome, to learn from those who are the best. The term was popularized in the early 1980s by Xerox, who, having fallen badly behind their competitors, used Benchmarking to regain competitive advantage.

Benchmarking is based upon the quality management concept "management by facts." Through it, an organization sets goals based upon what it has seen other organization do well to achieve best-in-class status. In the early 1990s, this tool was further enhanced by the formation of the Benchmarking Clearinghouse by the American Productivity and Quality Center (APQC) in Houston. The APQC has furthered and moved the state of art forward by establishing a process for Benchmarking project selection, and by designing methods to operationalize the process using teams and facilitators.

Some of the case studies and profiles found in this book originated from a conference held by the APQC in 1994. This conference also added a new dimension by bringing in the best in class to present and contrast Breakthrough strategies in order to see what works and what doesn't work.[5]

Benchmarking Defined

Benchmarking can be defined as the systematic comparison of elements of the performance of an organization against that of other organizations, usually with the aim of mutual improvement.[6] The Japanese word equivalent is "*dantotsu,*" which means "striving to be the best of the best." Each of these terms has been defined by Benchmarking expert Carl Thor as follows:

- **Systematic** means that the comparison is carefully planned, data are gathered and interpreted specifically for this purpose and organized improvement is a direct result of this process.
- **Elements of Performance** that are to be compared will vary from organization to organization. Thus, it is vital to know your own organization first.
- **Organization** refers to any size or type of organization, ranging from an entire company to a small work group. The most meaningful comparisons are often at the middle-level, functional sections of a department, or at the core process level that cut across an organization, such as order processing.
- **Mutual Improvement** excludes industrial espionage and market research related to competitive advantage. To get something, you must give something. Although the term competitive benchmarking is still around, most efforts are with non-competitive organizations. As a result, many of the heavily benchmarked issues are partly generic across industry lines.

Types of Benchmarking

There are three major types of Benchmarking efforts: (1) internal benchmarking, (2) strategic benchmarking and (3) generic cross-industry study. Of the three, the strategic focus is most useful to logistics practitioners in that it forms the basis of an industry study or survey. These are conducted by trade associations who specialize in a particular aspect of logistics, and who present their findings as general tendencies of industry segments or in a disguised format. Typical strategic benchmarking issues include:

- major equipment capacities, configurations, and problems
- personnel statistics and mixes
- customer satisfaction results
- maintenance practices and costs
- vertical integration of all organizations in the value chain
- internal quality data
- environmental practices

Steps to Effective Logistics Benchmarking

Because Benchmarking is a detailed discipline, it is necessary to follow a proscribed routine, even though some of the steps may be reordered or skipped:

1. **Choose the right logistics topic** — most Benchmarking projects wind up being mixed, even though they may start out with one topic, which in many cases is based upon executive intuition.
2. **Choose the Benchmarking team wisely** — the quality of the Benchmarking team will make or break the effort. Using the team approach minimizes the individual expert approach. The team should be a diagonal slice of the organization, with representatives from all relevant areas of the value chain having an interest in the topic.
3. **Gather external data** — focus on finding out what theater already exists in the public domain. Since organizations like to brag about their excellent experiences, there are quite a few well-documented stories already written, some of which are Profiles in this book.
4. **Gather internal data** — in order to understand the value chain completely insofar as the particular topic is concerned. Because procedures manuals are almost always outdated, it is important to interview the doers of the process and define its intra-organizational boundaries, inputs and outputs, owner responsibilities, measures, inspections conducted, improvements underway, etc.
5. **Select a logistics partner** — who is already in the position of doing something well and is recognized externally for it. It is important to identify what your organization can offer to the "partner" in return.
6. **Organize field Benchmarking** — in order to set up the ground rules for confidentiality, priorities, and the timing of visits.
7. **Conduct field Benchmarking** — starting with a meeting of the partner's team to compare flowcharts and process maps. A major reason for Benchmarking failures is that site visits are done too early in the process and team members aren't sure what they are looking for.
8. **Implement the improvement** — using the data collected from the field visits.
9. **Focus on the next round** — based upon the lessons learned.
10. **Measure and document the process** — by zeroing in on the various performance measures and translating them into effective standards for those that will be implemented.

Most companies conducting logistics Benchmarking analyses are able to obtain data on their own organizations and are able to get a large amount of good data on best-in-class companies. The problem usually is that these two sets of data are not consistent, e.g., if it is cost data, it may be based on different accounting systems or spread over different organizational entities.

The most complex and most crucial part of the Benchmarking process is to get the internal data and the external data on an "apples-to-apples" basis. In fact, organizations doing Benchmarking for the first time often misunderstand the most important part of the process; instead they focus on the difficulty of getting outside data. However, the reason they are having that difficulty is because they think they have to get the outside data in exactly the same form as their internal data, which is often impossible.

The Benchmarking process is still in its infancy. It represents a shortcut approach to identifying or illustrating CSFs that can be used as input into the organization's strategic plan formation. In using Macrologistics CSFs, the focus has to be on the time and place dimensions, not just on cost and productivity. In addition, departments and processes that are often overlooked can become the areas where learning occurs on how to improve existing assets. The logistics and purchasing areas are a good place to begin.

Common errors in Benchmarking are developing instant solutions using processes that have worked well somewhere else. Although the CSF Benchmark provides useful information, an exact replica of a CSF related process (e.g., customer service automation) may backfire when grafted onto a new environment. In the next chapter, we discuss the process reengineering technique that is used to operationalize the CSFs, after they are crystallized and defined.

References and Endnotes

1. The CSF concept was first presented in the March-April 1979 issue of *Harvard Business Review* in an article titled "Chief Executives Define Their Own Data Needs," and was expanded on in a series of technical reports by Rockard and are available through the Massachusetts Institute of Technology (MIT).
2. Spechler, Swendeman, and Voehl, 1980. Much of the description of critical success factors in this chapter was adapted from "Making Stategic Planning Come Alive," by Frank Voehl, *Strategy Digest*, 1993.
3. Rockart, 1979.
4. *Global Quality*, by Richard Tabor Greene, p. 149.
5. The APQC's Benchmark Clearinghouse has formally supported this book project by mailing out a letter of project endorsement to 20 or so companies participating in the conference, from which the Profiles contained in this book were prepared.

6. *The Measures of Success: Creating a High Performing Organization*, by Carl Thor, Oliver Wight Publications (Wiley), 1994.

General Electric: The Company of Tomorrow, Today — 9.1

General Electric is already the company of tomorrow. In a bold set of management actions aimed at increasing innovation to improve productivity to make the company more competitive, GE has set the trends that others will follow into the 21st century. CEO Jack Welch, long known for his strategic philosophy of buying and selling firms to gain the number 1 or number 2 position in any industry in which GE is a player, recognized that maintaining those lofty sales and profit positions depends on improving the management approaches used throughout the company. Welch aims to increase the flow of ideas from all employees by improving the way they are managed.

Three new tools form the core of this management revolution: the workout, best practices, and process mapping. In the **workout**, a manager and his or her subordinates gather for a three-day retreat. Subordinates work on problems with the help of an outside facilitator; the manager does not participate in these sessions. On the third day the manager is asked to respond to solutions proposed by subordinates with a yes, a no, or a deferral for further study. (Managers are encouraged to limit the number of deferrals.) What makes the workout intriguing is that the manager's own supervisor is present on the third day, but the manager is not allowed to witness his or her reactions to the employees' suggestions. The manager faces the group of employees, and the supervisor sits behind the manager, also facing the employees. The intent is to involve employees in decision making, to solve problems, and to change managers' attitudes toward employee involvement.

In the **best-practices** technique, the firm compares itself with the firms that are best at performing a particular function or process. GE, for example, has compared itself with the best firms in making appliances and lightbulbs and in performing various financial functions, such as making loans. Once the comparison has been made, GE attempts to improve its performance levels by emulating the best practices of other firms. Significant improvements have been reported throughout GE's many businesses as a result of best-practices analyses.

In **process mapping**, employees complete a flowchart of a process such as making a jet engine. The flowcharts show how all the component tasks are interrelated. Employees then try to see how much time they can cut from the process. In the case of jet engines, which GE has been making for years, the firm was able to cut the manufacturing time in half through process mapping, thereby saving many millions of dollars.

What makes all these approaches work is a change in GE's corporate culture, which has become obsessed with productivity and innovation. Welch has been actively involved in the change by, among other things, attending the sessions in which the new programs were developed. He has laid out the changes he wants and has offered rewards for success. Ever ready to experiment and make things happen, Welch is confidently leading GE into the 21st century.

© James M. Higgins. *Source:* Thomas A. Stewart, "GE Keeps Those Ideas Coming," *Fortune* (August 12, 1991), pp. 41–49.

Motorola Invests in Its Future 9.2

Recognizing the various strategic challenges it faces between now and the turn of the century, Motorola has decided that "the most crucial weapons will be responsiveness, adaptability, and creativity." To acquire these organizational skills, Motorola has initiated a lifelong-learning program for its employees. The program, established by former chairman Robert W. Galvin, is designed to provide training and development for every employee, from the shop floor to the top echelons.

The financial commitment to the program is significant, not only in terms of the cost of training and development, but also in terms of time lost from work. Motorola already gives every employee at least forty hours of training and development a year, spending an amount equal to 4% of sales a year on training and development. It expects to quadruple those levels by the year 2000. The expected cost is $600 million a year — about the cost of a new chip factory. CEO Gary L. Tooker comments "If knowledge is becoming antiquated at a faster rate, we have no choice but to spend on education. How can that not be a competitive weapon?"

Much of the company's willingness to invest in education comes from its past successes with training and development. Motorola University, headquartered in Schaumburg, Illinois, has fourteen branches in locations from Tokyo to Honolulu, and a budget of more than $120 million. Courses are designed by "instructional engineers" and cover such diverse topics as critical thinking and problem-solving management, robotics and computers, and remedial English. Courses are specifically designed for each geographic area so that cultural references are relevant. Motorola also works closely with its suppliers to train them in ways that benefit both parties — for example, by providing training in problem solving and total quality management.

One of the key areas for improvement in the future is innovation and creativity that leads to it. Some of the corporate skills that have made Motorola a world-class competitor also make it rather inflexible and thus unable to innovate well. In particular, the degree of regimentation required if a firm is to be a high-quality, low-cost competitor tends to stifle innovation. Fortunately, Motorola has a culture that encourages conflict and leads to continual improvement. Moreover, Motorola's culture encourages the pursuit of neglected technologies and uses teams to help it innovate and anticipate change. It is among the top then U.S. firms in absolute annual spending on R&D.

Motorola, now the global leader in cellular phones, pagers, two-way radios, and microchips (used to control devices other than computers), recognizes that to remain on top of tomorrow's business situation it needs to make some changes. It has already begun to develop in its employees the independent thinking that is vital to innovation. Yet much remains to be done.

© James M. Higgins. *Sources:* Kevin Kelly and Peter Burrows, "Motorola: Training for the Millennium," *Business Week* (March 28, 1994), pp. 158–163; G. Christian Hill and Ken Yamada, "Staying Power: Motorola Illustrates How an Aged Giant Can Remain Vibrant," *Wall Street Journal* (December 9, 1992), pp. A1, A18.

Heroes on the Help Desk: Redefining the Strategic Role of Support — 9.3

The principals of Renaissance Partners, Inc. have put together a provocative story of how effective Information Technology support services are saving their client organizations millions and improving customer satisfaction at the same time. Although Help Desks have been around for almost twenty years assisting customers and internal staff with PC and other technical problems, not everybody is happy with the concept. As computers have made their way from the glass palace to the desktop, the cost of supporting the infrastructure has risen faster and higher than anyone expected, contend the authors. Global expenditures in 1995 for technical services and support have exceeded $175 billion and the percentage of the IT budget that goes to support services, rather than to operations and development, averages 57% and is climbing.

Clearly, a new approach is required, one that strategically redefines and repositions the Help Desk to sit astride not one but three critical processes: Contacts, Incidents, and Problems. Thus, the function must be reengineered to best satisfy these three processes and a number of companies are profiled and their attempts documented:

- **Clorox** reengineered IT support services to create measurable strategic value and short term tactical results, with a goal of reducing the hidden cost of IS/IT support services by 50%.
- **Florida Power and Light** developed a training program to familiarize employees with basic PC/LAN concepts, resulting in a reduction of calls by some two and one-half times.
- **IBM** created a special support service for external customers to provide effective Help Desk functions such as assessment, automation services, integration support, and operations assistance.
- **Vantive Corporation** is the fastest growing, publicly traded software company in the customer-interaction market and the leading provider of integrated customer interaction software covering automated sales, marketing, customer service, defect tracking, field service, and internal Help Desk functions.
- **Software Artistry, Inc.** develops and markets a product called SA-Expertise which provide the enterprise support management tools required to link operations such as Help Desk, Network Management, Asset-Change Management, and End-User Empowerment.
- **Clarify, Inc.** provides systems to assist clients with product support call management, Help Desk management, product change requests, and inventory control.
- **Entergy Corp.** used the Quality Action Teams approach, coupled with natural work teams, to completely reengineer its 13 Help Desk operations in three states into a single "Command Theater" operation and Double the number of calls processed with a 15% reduction in staff.
- **Taco Bell** used the SCORE system to separate the Help Desk support from the IT group in order to become more focused on the business of supporting restaurants.
- **Magic Solutions, Inc.** have worked with over 27,000 organizations worldwide to improve Help Desk operations, including the White House and were recently rated #1 by Software Digest. They feature imbedded artificial intelligence to assist the help desk professionals in solving complex problems and shorten the support cycle.
- **Remedy Corporation** has developed the Action Request System to help track and resolve support requests in the over 2,000 PC and UNIX computing environments in 35 countries involving some 1,000 customers.
- **NASCO** used the Bendata HEAT software system to centralize and revamp its Help Desk operations, including the education of customers on tools, services, correct usage and options available.

Source: Steve Murtagh and R. William Sheehan, *Fortune*, April 15, 1996, pp. 69–85; Renaissance Partners, Inc., SMDT & Report, 1996.

American Productivity and Quality Center (APQC): Research and a Lot More 9.4

The APQC is a Macrologistics-based quality research center that accumulates information on best practices for all types of processes. It performs client research on who has the best practices, supplies copies of articles on processes, conducts seminars, and sells publications. Founded by Jackson Grayson in the late 1970s, it has its roots in the wage and price controls movement under President Carter and others. It has become the leading center for Benchmarking activities in North America under the direction of the Benchmarking Clearinghouse and its president, Dr. Carla O'Dell.

In addition to its quarterly publication, *Continuous Journey,* the APQC offers the following information-based quality management services:

- Targeted research, comprehensive studies, and in-depth investigations.
- Knowledge management, Total Quality Management, and change management.
- Studies covering best practices, business process innovators, potential Benchmarking partners, and reengineering.
- Strategic planning, human resource management, quality strategy, critical success factors, and key learnings.
- Unique sources, business solutions, and skilled professional researchers.

A recent Macrologistics tool developed jointly by the APQC and Arthur Anderson is the Knowledge Management Assessment Tool (KMAT). This innovative tool proposes how four organizational enablers — leadership, culture, technology, and measurement — can be used to nurture the development of organizational knowledge through the Knowledge Management Process. The criteria in the KMAT enables an organization to assess how important each of these criteria is to their organization, and how it is performing in relation to the criteria, ultimately coming up with a final "score." This score will reveal strengths and weaknesses, as well as individual directions for improvements, according to APQC Chairman Grayson.

The APQC is located in Houston and can be reached at the following address: 123 North Post Oak Lane, Houston, Texas, 77024; Tel (713) 681-4020; Fax (713) 681-5321; e-mail apqcinfo@apqc.org and at web site http://www.apqc.org. In addition, the center operates a proprietary electronic network that links over 400 APQC members.

Source: American Productivity and Quality Center, Houston. We are indebted to them for their encouragement and sponsorship during the early stages of this work.

INTEGRATION

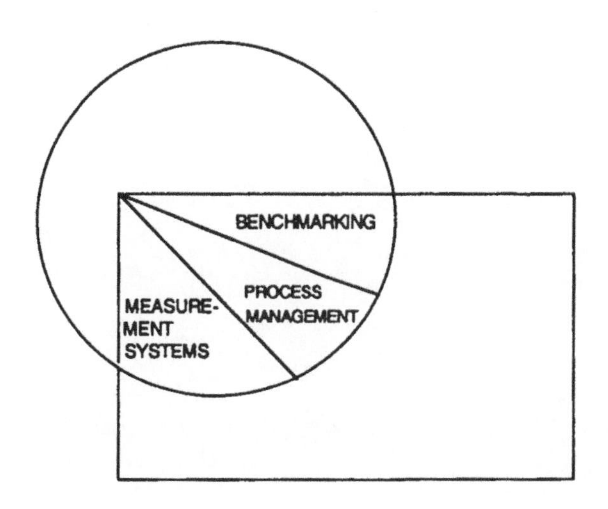

10 Process Management and Process Reengineering

Process Management represents an operational way of acting upon CSFs and the Macrologistics strategy. This tool is critical to starting a change effort and any attempt to streamline and improve the value chain will involve some form of Process Management. Achieving Breakthrough strategies often involves the use of a well-planned process reengineering effort, which often works well with adaptive engineering.

> "A Process is a structured, measured set of activities designed to produce a specific output for a particular customer or market... A process is thus a specific ordering of work activities across time and space, with a beginning, an end, and clearly defined inputs and outputs: a structure for action."
>
> Thomas Davenport

Process Defined

A process is a set of linked activities that take an output and transform it into an input. According to Johansson and other process experts,[1] there are two ways to define and interpret the notion of a business process among the various practitioners in the logistics field and related quality movement. One definition is a process is a workflow, having a series of activities aimed at producing something of value. The other view of process is as the coordination of work, whereby a set of skills and work routines are exploited to create a capability that cannot be easily matched by others. The approach gets further complicated by the approach taken, whether it be evolutionary or revolutionary. Business Process Management is an integrated combination of these four choices to allow maximum flexibility to shift from one to another as conditions change, using Process Mapping as the integrative tool.

Process Management

A comprehensive definition of Process Management would include the following key concepts:

- It is the organization of work to achieve a result.
- It involves multiple steps and coordination of people and information.
- Its design creates a distinctive competitive asset for the organization in areas of research, product development, process execution, and core competencies.
- It establishes management as the enabler and sustainer of process advantage.

Process Mapping is a tool used to help management and workers gain a fresh process insight. When applied to the logistics process, it may help identify areas where breakthrough is possible. Traditional thinking about processes comes from "process blindness."[2] Because managers have not had to pay much attention to processes, this blindness has plagued most companies for many years. A 1994 study that appeared in Information Week cited that on an operational level, most senior managers have no idea how their companies operate. In other words, real day-to-day operational performance is no longer understood, nor is it controlled on a real-time basis.

Managers focused on strategy, planning, organization, and market/product innovation. The process movement has now forced these managers to take a fresh look at their firm's business processes in order to make business process investment their own responsibility. And to do this, they need sources on insight. And one of the greatest practical hurdles to developing process insight is the business mindset that operates in terms of functions. Also, if competitors do something differently and help create new Critical Success Factors (CFSs), the process may need to be changed to keep pace and break away from the old views.

In order to break away from the traditional view where single functions dominate as a natural way of thinking, Keen and Knapp pose several questions for logistics management to answer:[3]

1. Who exactly is the customer or person to which the outcome matters?
2. What must happen for the customer's request to be completely satisfied?
3. Who does the work and how does it come together?
4. How will the work be coordinated logistically?
5. Can Information Technology be exploited to improve coordination? Empower the people to do the work? Augment training? Alter incentives?

The answer to the above questions marry the workflow and coordination activities. They look at process, not function or activity. They do not prejudge the solution. Process Management allows for either an evolutionary approach of continuous improvement or a revolutionary one such as reengineering. Also, some processes such as logistics cannot be reengineered quickly and thus, there are barriers to meeting customers needs. By using evolutionary methods such as adaptive engineering, there is a possibility for rapid and significant change that delivers a breakthrough, by integrating existing processes that have been adaptively modified.

This approach advocates that to be truly effective and deliver long-lasting results, the way to improve the way people work must be evolutionary, not revolutionary, in most cases. The long history of TQM failures, coupled with the broken promises of reengineering, suggest that evolutionary practices deliver more permanent results. It stimulates morale and imagination, creating conditions for rewarding organizational learning. This approach also inspires employees to discover innovative ways to deal with adversity and competitive challenge.[4]

Business Process Reengineering

Reengineering is by far the most influential and most controversial interventions within the process movement. It calls for a radical change in the business process as a matter of survival for companies and targets dysfunctional or broken, outdated processes for investment and redesign. The redesign effort usually requires a major investment in information systems and technology.

As defined by its leading proponents, Michael Hammer and James Champy, it is the fundamental rethinking and radical redesign of business processes to achieve dramatic breakthrough improvements in critical, contemporary measures of performance, such as cost, quality, service, and speed.[5] While traditional quality approaches start with stable processes and the need to improve them (or first stabilize them), reengineering throws away how things are done today, and starts over with a blank slate to design new processes and organizational structures that will achieve, in one fell swoop, breakthrough and competitive advantage.

Reengineering calls for starting with a clean sheet of paper and asks that if we were to start this company today, how would we design the process? It assumes a fundamental disconnect between a company's current course and what is required for success in the marketplace. As such, it will be required both of companies that never got off the quality bandwagon, as well as those whose markets are becoming increasingly more turbulent and are changing faster than their current logistics programs have been able to handle.

The major process opportunity lies in the cross-functional streamlining of work activities. It attacks division of labor and functional organization of work as the biggest constraint in service improvement and higher. The credo is "don't automate — obliterate!" Most organizations that attempt to implement reengineering settle for a form of Business Process Redesign instead due to questions concerning ethics, practicality and necessity.[6] These limited scope activities are also not likely to deliver breakthrough changes because they focus on subsystems of functions, which are rarely cross-functional in scope. We have found that the total delivery system must be put under the microscope if reengineering is to succeed and significant changes are made in the value chain. What makes the reengineering process so powerful is the blending of the various components into a synergistic whole.

Process Salience and the Process Paradox

Viewing processes as workflows tells us something about their objectives and interfaces, but in no way indicates their relative importance to an organization's strategic vision or its competitive position. Process Management has at its core the notion of process investment: the salience and worth of a process. Salience refers to the role of the process, while worth looks at the asset value. The salience of a process can be defined in terms of its attributes[7] or conditions in the business process management life-cycle:

1. **Identity processes** define an organization to itself, its customers, and its investors.
2. **Priority processes** are the engines of everyday work performance.
3. **Background processes** are essential to operation but do not contribute to strategic success.
4. **Mandated processes** are those imposed upon an organization by regulatory agencies.
5. **Folklore processes** are those carried on long after their useful life is over.
6. **Macrologistics processes** are those that can significantly transform the time and service attributes of an organization to deliver a substantial competitive advantage.

A logistics process paradox states, however, that the immense benefits do not directly translate into business value. An estimated total of 50 to 75% of all reengineering projects have failed, which is comparable to the TQM failures of the early 1990s. This paradox is summarized in a 1993 report by McKinsey and Company on business process reengineering in 100 companies:

> In all too many companies, reengineering has not only been a great success but also a great failure. After months, even years, of careful redesign, these companies achieve dramatic improvements in individual processes only to watch overall results decline... managers proclaim a 20% cost reduction, and a 25% quality improvement — yet in the same period business unit costs increase and profits decline.[8]

So we must ask the question: is this failure? Or does the problem lie with the process itself, which in many cases is never really implemented. In other words, they talk about it but rarely actually do it. In Macrologistics Management, reengineering evolves into a form of "organizational reengineering:" a fundamental rethinking of operating processes and organization structure, focused on the organization's core competencies to achieve dramatic improvements in organizational performance.[9] In other words, it is a process by which any organization can redesign the way it does business to maximize its core competencies, of which logistics competency is critical to success.

Organizational reengineering redesigns the way work flows through an organization, often leading to system and infrastructure changes. It differs from traditional reengineering by focusing on the core competencies, or CSFs for success, and are tied together logistically in a four-fold manner:

- A greater focus on the organization's customers.
- A fundamental rethinking of the logistics processes that lead to value-added improvements in productivity and cycle time.
- A structural reorganization, which typically breaks functional hierarchies into cross-functional teams.
- New information systems and measurement systems, which use the latest in technology to drive improved data distribution and decision making.[10]

The process paradox was at work in IBM and General Motors at a time when these companies were awarded the Malcom Baldrige National Quality Award. Their economic and competitive performance was plummeting while the trophies were flowing. The problem was that, among other things, these companies did not fully **integrate** the process out into the value stream where long lasting change could be accomplished. They failed to learn quickly enough from the units where process change started and succeeded, nor were they able to change system level processes across organizational units. While plant level change may win an award, it is unlikely to have a significant competitive impact.

Logistics Process Enhancement

Logistics process enhancements can be achieved in the following three common process areas which flow throughout the entire value chain: the demand process, the supply process and the delivery process.

It is necessary to focus on the demand process first because if the organization doesn't understand what it has to deliver, and it doesn't fully appreciate the force of the demand-pull influence, then improvements or reengineering efforts in the subsequent process would not matter much. As a "core" process, demand has the following sub-processes associated with it: Target Stocking Levels, Back Order management of open orders, and Level-of-Service vs. Level-of-Availability monitoring.

To attack the supply core process next, a three-pronged approach is recommended to:

- Establish the correct replacement models
- Ensure that vendor management correctly aligns with customer requirements

- Implement and validate a defective item identification and repair cycle for efficiency and timeliness to minimize process inventory and strengthen competitiveness.

Once the demand and supply core processes are addressed, the delivery process needs to be addressed. This core process has three sub-processes involved in the improvement effort: resource sharing, replenishment, and regional stocking location (RSV)/distribution center (DC) network.

To effectively analyze these core and sub-processes involved in the logistics model, the process mapping technique can best provide process insight while identifying the chief characteristics for process improvements. The Process Mapping technique can also be used to test a process for investment capability based upon an analysis of process salience, once the question of whether to organize around processes is settled.

Reengineering and Whole Systems Change[11]

Like some effective reengineering efforts, Whole Systems Change produces an easily understood payoff by almost immediately producing positive changes on the organizations bottom-line. This is most often a result of corporate downsizing via restructuring, as well as the impact upon the supply chain of the organization, as outlined in Chapter 8. In addition, reengineering using the Whole Systems model, has the positive effect of changing a firm's work orientation from bureaucratic, political and disjointed to one that is more synchronized, uniform and efficient (Childress & Senn, 1995).

Some claim reengineering has too much "top-down" orientation, planning and execution and, as a result, it fails to fully consider all elements of the organizational system (such as customers or suppliers). By integrating reengineering with Whole Systems change, using the Whole Organizational System approach, breakthrough results can be achieved.

The MaxComm Whole Organizational Systems Approach

A number of organizational theorists feel the reengineering approach fails to adequately emphasize *where* the organization wishes to go with its transformation effort and *who* is involved in getting it there. The Whole Organizational Systems approach places primacy upon addressing these issues and was developed by MaxComm Associates to facilitate organizational transformations in client organizations.

It focuses much less on a structured, measured process than some other change approaches (like reengineering) and much more on creating an environment that promotes flexible, learning systems that will close the "future gap."

The term Whole Organizational Systems describes the scope of the change effort: it attempts to involve representatives of all the organization's systems — executives, union personnel, hourly personnel and even customers and suppliers — both in the identification of the ideal future state and the change effort used to get there.

At its core, Whole Organizational Systems change presumes that an organization can effect transformation by giving people control over the economies and technologies of their work.

The heart of Whole Organizational Systems change is the *future search conference*: an organization-wide meeting in which participants identify how their organization would look in the ideal future and how their organization looks in its present and historical perspectives. This reveals a "gap" between where the organization is and where participants want it to be — allowing individuals to, in essence, plan the organizational change *backwards* to desirable goals.

The future search conference is based on three assumptions:

- That change is so rapid that an organization needs increased face-to-face communication to make intelligent decisions.
- Successful strategies come from envisioning preferred futures.
- People display more commitment to plans they help to develop.

Using this structure in tandem with the established vision of the future and teamwork principles, teams across the organization utilize process redesign to implement the change effort's action plans.

The Whole Organizational Systems change process must first begin with a predetermined effort to clearly define the need for change. The advantages that result from this approach include.

- It involves the "Whole-Organizational System" from employees to customers.
- Begins with an ideal future state and then plans *backward* to achieve that state.

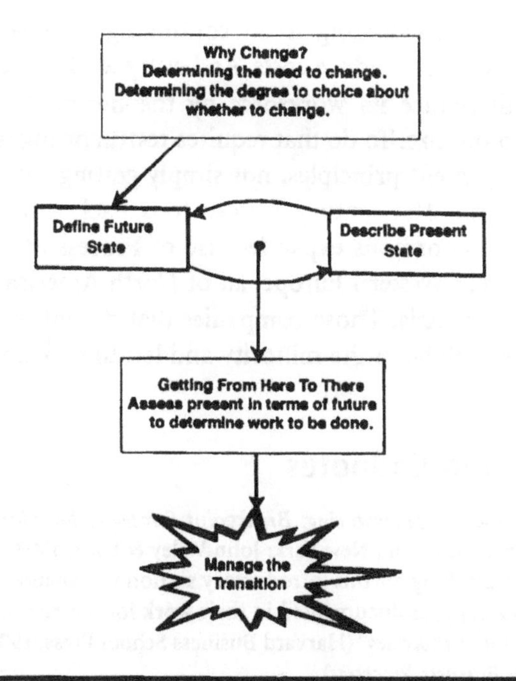

Figure 10.1 The MaxComm Whole Organization Systems Approach

- The fundamental premises of the approach inherently foster a flexible, empowered environment that can better respond to market needs and improve employee satisfaction.

Summary

Many of these Process Management approaches are relatively recent additions. As more implementation of change comes from group conferencing rather than structured executive planning as in reengineering, the implementation of change may easily become sidetracked. These approaches are heavily reliant upon the effective teamwork and leadership skills of organizational members. Without appropriate logistics management training and ability in these skills, organizational members may simply be unable to carry out the tasks asked of them. At the same time, work is becoming more mobile as workers, customers, and offices move around in cars and through electronics. Information Technology is being applied in more ways than ever before.

We also see the downsizing trend continuing as organizations flatten hierarchies in an effort to reduce costs. And every week it seems that a major corporation will reduce its workforce by the hundreds and thousands, depending upon the size. To do that requires restructuring work using Macrologistics Management principles, not simply getting rid of it. Perched at the threshold of the 21st century, we can look back with amazement and satisfaction at the enormous expanded use of Process Management in the logistics area, across Western Europe, all of North America, and the industrialized portions of Asia. Those companies that do not seriously integrate these approaches will be in the minority and left to pick up the scraps.

References and Endnotes

1. *Business Process Reengineering: Breakpoint Strategies for Market Dominance*, by Henry Johansson et al., New York: John Wiley & Sons, 1993.
2. Part of the thinking for this introductory section was adapted from the works of Keen and Knapp, as documented in their work for the Harvard Business School titled *"Business Processes"* (Harvard Business School Press, 1996, Section 1, *Introduction to Business Processes*).
3. Ibid., pp. 18.
4. *The Hocus-Pocus of Reengineering*. by Paul Strassmann, Across the Board, Conference Board of New York , June, 1994. pp. 33–34.
5. Michael Hammer and James Champy, *Reengineering the Corporation: A Manifesto for Business Revolution* (New York: Harper Business, 1993) p. 32.
6. See *"Process Innovation: Reengineering Work Through Information Technology"*, by Thomas Davenport, Harvard Business School Press, 1993.
7. *Business Processes*, by Peter Keen and Ellen Kamp, Harvard Business School Press, 1996, pp. 30–33. Also see "Process Management" by Bill Meyer and Frank Voehl, New Florida Technologies, 1997 and Strategy Digest, 1998.
8. *"How to Make Reengineering Really Work,"* by Gene Hall, Jim Rosenthal, and Judy Wade. Harvard Business Review, November-December, 1993, pp. 119.
9. For further discussion, see *Reengineering the Organization: A Step-by-step Approach to Corporate Revitalization*, ASQC Press, 1994.
10. Ibid., p. 421 (As reported in the 1995 *Quality Yearbook*, McGraw Hill).
11. This section is based upon the work of MaxComm Associates and their CEO Bill Adams, on Whole System Change Management, MaxComm Associates, 1996.. See Process Management and Whole Systems Change, by W.A. (Bill) Adams and Frank Voehl, *Quality Observer*, April, 1998; *The Whole Systems Change Manual*, MaxComm Associates, Salt Lake City, Utah, 1973..

Xerox: Duplicating Its Past Successes 10.1

In its heyday, around 1970, Xerox controlled 95% of the U.S. copier market. In 1982, however, it had only a 13% market share. Its top executives had fallen asleep at the switch. In the late 1980s CEO David Kearns wrested control of Xerox's culture and turned it into what many considered to be an American samurai. Kearns directed the effort that enabled Xerox to win Japan's Deming Prize, the United State's Baldrige Award, and the European Quality Award. Xerox is the only firm ever to have captured all three of these awards.

But quality wasn't the only problem; innovation was another major issue. Xerox's brilliant scientists were turning out good ideas, but the corporation wasn't taking those ideas to market. Although in the early 1980s Xerox had invented the user-friendly PC as we know it today and the PC mouse, its management decided to make higher-priced copiers instead. Critics note that Xerox missed one opportunity after another.

Kearns's leadership brought Xerox to the point where it could compete on the basis of both quality and price in all but the low-end market. Kearns made Xerox's core business competitive again. His successor, Paul Allaire, has formulated a strategy aimed at positioning Xerox for a world that is changing from analog to digital information storage, retrieval, and exchange. Allaire comments, "We expect to be the preeminent company in the document field." He wants Xerox to be "The Document Company." A critical component in the implementation of this strategy will be innovation, especially product innovation.

To achieve his strategy Allaire has moved to align Xerox's Seven S's so as to achieve the company's goals. Although it hasn't been easy, financial analysts agree that the arrows are all pointing in the same direction (see Chapter 4). Allaire points out that "A lot of times people will just change the structure and reporting relationships. But if you want to change a company, you'd better change more than that. There's the formal structure and then there's the way the company really works. You have to change the way it really works." Among the many actions he has taken to achieve this goal are the following:

- Increasing the number of strategic alliances in which Xerox participates, recognizing that the company cannot do everything by itself.
- Switching from a functional to a business unit structure, with each semiautonomous unit based on the final customer. Each business unit is accountable for its profits and losses, and each has its own manufacturing and sales group, as well as allocated research and service teams.

- Developing a closer relationship between Xerox's renowned research and development unit, PARC (Palo Alto Research Center), headed by Chief Scientist John Seely Brown, and its marketing and manufacturing functions.
- Creating a bonus system based on individual, unit and corporate performance, thus tying each employee's future more tightly to the company's performance.
- Empowering employees.
- Downsizing.
- Changing the corporate culture to focus not only on quality and cost but also in innovation.
- Creating new-venture spinoffs when business units do not feel that a specific research result matches their product lines.
- Embracing change. Allaire asserts, "We will need to change Xerox more in the next five years than we have in the past ten."
- Introducing several major products. One of these, DocuTech, is extremely versatile — it can scan, copy, print, and even bind and staple small booklets. Using DocuTech, Blue Cross/Blue Shield of North Carolina found that it could put together benefits booklets in two to three days; this activity formerly required six to nine weeks. DocuSP is a software architecture that works with DocuTech. Introduced in April 1994, it allows for data from all types of PC's located anywhere in the world to be input into a document capture machine. It then provides common frameworks for text, art, and graphics, which can be printed by a DocuTech system.

Peter van Cuylenberg, Executive VP, responsible for Xerox's new emphasis on digital processes, describes what the company has accomplished: "What is new here is that we've looked at the whole document pipeline, from creation to use, and provided all of the software and hardware to build the whole infrastructure. No one has been able to integrate the system we're talking about because the components didn't exist until now."

 Source: Subrata N. Chakravarty, "Back in Focus," *Forbes* (June 6 1994), pp. 72–76; Tim Smart, "Can Xerox Duplicate Its Glory Days?" *Business Week* (October 4, 1993), pp. 56–58.

Xerox's Systematic Problem-Solving Process Assists Their Process Management Effort

Systematic problem solving in most firms draws heavily upon the philosophy and methods of the quality movement. Real learning demands hard facts to support problem solving; otherwise, it cannot take place. Thus, firms that practice organizational learning use the following techniques:

- The scientific method, as opposed to guesswork, for recognizing and identifying problems (what Deming referred to as the "Plan, Do, Check, Act" cycle and others call hypothesis-generation, hypothesis-testing techniques).
- Data, rather than assumptions, for solving problems. This is referred to as "fact-based management."
- Simple statistical tools (histograms, Pareto charts, correlations, cause-and-effect diagrams) to organize data and draw inferences.
- Simple idea generation techniques, such as brainstorming, for generating alternative solutions.

In 1983, Xerox launched its Leadership Through Quality initiative. Since then, virtually all Xerox employees have been trained in small-group dynamics and problem-solving techniques. They follow the six-step problem-solving process shown in the following table.

Employees have been provided with problem-solving tools in four areas: collecting information and generating ideas (interviewing, surveying, and brainstorming); reaching consensus (rating forms, list reduction, weighted voting); analyzing and displaying data (force-field analysis and cause-and-effect diagrams); and planning actions (Gantt charts and flow charts). The result has been a consistent, scientific approach to problem solving. Employees are expected to use these techniques in all meetings, and no topic is off limits. CEO Paul Allaire states that when a group was formed to review the firm's organization structure and suggest alternatives, it used the same process.

To determine the effectiveness of what Xerox has done, look at its track record since it became a quality learning organization. It has recaptured market share, raised its quality to high levels, created many new products, and cut its costs.

 Sources: David A. Garvin, "Building a Learning Organization," *Harvard Business Review* (July-August 1993, pp. 81–82; Robert Howard, "The CEO as Organizational Architect: An Interview with Xerox's Paul Allaire," *Harvard Business Review* (September-October 1992), p. 106.

XEROX's Problem-Solving Process

	Step	*Question to Be Answered*	*Expansion/Divergence*	*Contraction/Convergence*	*What's Needed to Go to the Next Step*
1.	Identify and select problem.	What do we want to change?	Lots of problems for consideration.	One problem statement, one "desired state" agreed upon.	Identification of the gap "desired state" described in observable terms.
2.	Analyze problem.	What's preventing us from reaching the "desired state"?	Lots of potential causes identified	Key cause(s) identified and verified.	Key cause(s) documented and ranked.
3.	Generate potential solutions.	How could we make the change?	Lots of ideas on how to solve the problem.	Potential solutions clarified.	Solutions list.
4.	Select and plan the solution.	What's the best way to do it?	Lots of criteria for evaluating potential solutions. Lots of ideas on how to implement and evaluate the selected solution.	Criteria to use for evaluating solution; agreed upon. Implementation and evaluation plans agreed upon.	Plan for making and monitoring the change; measurement criteria to evaluate effectiveness of solution.
5.	Implement the solution.	Are we following the plan?		Implementation of agreed upon contingency plans (if necessary).	Solution in place.
6.	Evaluate the solution.	How well did it work?		Effectiveness of solution agreed upon Continuing problems (if any) identified.	Verification that the problem is solved, or agreement to address continuing problems.

Source: David A. Garvin, "Building a Learning Organization," *Harvard Business Review*, July–August 1993. Based on a chart in *Innovate or Evaporate*, by James Higgins originally published a a table for building a learning organization, by James Garvin, Harvard Business Press, 1993.

IBM Restructures in Order to Reinvent Itself 10.3

In 1988, CEO John A. Akers enacted a major reorganization. He took control of the firm away from a six-person management committee and reorganized the firm into six strategic business units. That reorganization was met with considerable positive customer response. In December 1991, Akers announced the second major restructuring in three years. He planned to divide the firm into thirteen autonomous units, nine manufacturing and development lines of business, and four geographically based marketing and services companies to sell what the nine manufacturing units produced.

The three largest manufacturing and development businesses were Enterprise Systems which made mainframes, related processors and software; Adstar which made storage devices, tape drives and related software; and Personal Systems which made PC's, workstations, and related software. The four geographic sales units were: IBM Europe/Middle East/Africa, IBM North America, IBM Asia Pacific, and IBM Latin America. Critics heralded this reorganization as long overdue. It was generally believed that the increased autonomy would lead to greater market responsiveness and increased product innovation. IBM's bureaucracy had long been recognized as thwarting innovations. Its mainframers had managed to stave off those who wanted to move ahead with smaller, ever more powerful personal computers, workstations and client servers for fear of cannibalizing mainframe sales. The reality of course was that competitors did it for them. This restructuring was accompanied by a major downsizing. Including the cuts announced at this time, Akers had reduced IBM's work force from 407,000 in 1986 to 302,000 at the beginning of 1992.

In April, 1992, Louis Gerstner came to IBM from RJR Nabisco. Seven months later, he scrapped the plan to break the firm into 13 separate companies. He had discovered two sets of compelling reasons for doing so. Externally, several thousand customers that Gerstner had talked to did not want the company broken up into separate companies. Rather, they wanted IBM to deliver an overarching, cohesive perspective that solved their business problems, as opposed to the myriad of vendors with distinct products that couldn't function together. Internally, the firm was about to launch the most pervasive strategic technology thrust in its history. Gerstner's actions were taken because he believed that this new technology, labeled "Power," could best be sold through a unified firm. In his actions, Gerstner clearly sought a structure that would support the firm's innovation strategy.

 Sources: David Kirkpatrick, "Gerstner's New Vision for IBM," *Fortune* (November 15, 1993; pp. 119–126; Joel Dreyfuss, "Reinventing IBM," *Fortune* (August 14, 1989), pp. 36–39; Michael W. Miller, "IBM's Customers Know About Problems Akers is Dealing with in Reorganization," *Wall Street Journal* (February 1, 1988), p. 14.

11 The Balanced Scorecard Corporate Measurement System

The order process is the most common application of measurement systems to Macrologistics. The logistics cycle is concerned with the network of activities from when the customer places the order to when the delivered product has created the expected value and satisfaction to the customer. Thus, the order process is a major cross-functional process of the highest priority to the customer, but it is often unmanaged from the viewpoint of the total picture. A balanced scorecard measurement system is an important part of the solution to this problem. This approach doesn't build a massive new information system. Instead, it uses a streamlined approach to developing key indicators around critical success factors and business objectives.

"If you can't measure it, you can't manage it."

Peter Drucker

Corporate Measurement System Overview

While many organizations are embarking on some form of a logistics management program, few have implemented a measurement system that can be used to figure out how good of job is being done in the order process, as well as in the whole system of Macrologistics. From the studies conducted on the practices of successful organizations, however, some operating models have resulted. One of the most useful is a variation of the Corporate Measurement System (CMS) approach, based on the work of Jack Rockart of MIT, and Robert Kaplan and David Norton of the Harvard Business School. This model, called the Balanced Scorecard, suggests the use of a vital few Critical Success Factors, that link to the business objectives and processes, around which there are a group of balanced measures that summarize progress toward the objectives that are most important to the organization; thus the term "Balanced Scorecard".

The requirements of the logistic system which is intimately related to the order process, do not change from one area to another. Instead, they normally change from some areas to all areas. The logistics system must be both productive and customer oriented, both cost effective and flexible. To focus on service and quality is not enough and other aspects should also be included in the measurement system, which includes at least the following important measurement perspectives:

- Financial perspective (How do we look to our shareholders?)
- Customer perspective (How do our customers see us?)
- Internal Business Perspective (What must we excel at?)
- Innovation and learning perspective (Can we continue to create value?)

Each of these perspectives implies a set of goals that in turn link measures of performance to reach those goals. What organizations need is a "Balanced Scorecard" system that is simple and flexible in design, easy to use and modify, and is integrated in to key functions and processes. The information provided needs to be timely and accurate and must be perceived by the employees as truly useful and not just another "Big-Brother-is-Watching"–type system. Instead, what is needed is a measurement system that people can use to better manage their efforts and to link all areas of the company's supply chain to the corporate vision. However, a measurement system cannot be copied from one company to another. What can be used everywhere are procedures, attributes, methods, software, and rules for developing measurement systems. Therefore, a measurement system is both the measures themselves and the procedures for their use.

The following are the attributes that are needed in an effective logistics measurement system:

- Top management interest and support.
- Simple system that is easy to understand.
- Accurate, reliable information linked to strategy and business objectives.
- Good lines of communication.
- Specific objectives, procedures, and guidelines for use.
- Consistent, continuous monitoring.
- Assignment of specific responsibility and accountability.
- Timeliness of information.
- Competent monitoring staff who provide cogent analysis about relevant actions and interactions.
- Employee commitment and motivation to use.
- Periodic reports as well as remote terminal access.
- Actionable and practical information with linkages directly related to decision-making and fixing problems.

Measurement System Principles and Objectives

There are two guiding principles to be followed when developing Macrologistics measurement systems: (1) people in the supply lines of the organization respond best to information relevant to their piece of the world; and (2) when people have relevant information about things they deem important and can influence, they become very committed to using the information.

The following is a summary of the measurement system objectives:

- Translate the vision to measurable outcomes all logistics staff can understand.
- Focus and align the direction of staff based on measurable results.
- Track systems-related breakthroughs and continuous improvement results in the Value Chain.
- Foster accountability and commitment.
- Integrate strategic plans, business plans, quality and Benchmarking.
- Provide standards for Benchmarking.
- Problem solve business problems throughout the Value Chain.

- Provide basis for reward and recognition.
- Create individual and shared views of performance.

Types of Measures

There are three types of measures that must be considered when implementing a Logistics-Based Measurement System: (1) outcome (macro) measures, (2) Just-In-Time process (micro) measures and (3) Up-Stream Control (predictive) measures. Outcome measures are often called macro due to their broad nature that generally reflects an after-the-fact type of indicator. Examples are return on investment, or equity, overall customer satisfaction, program/project savings, etc. Micro (process) measures represent work-in-process types of situations and are often used for stopping the project, product, or program when problems, roadblocks, or rejects occur. The third kind, predictive measures, are used for "upstream control" or prevention-of-problem situations. Most effective measurement systems have an effective balance of macro, micro, and predictive indicators.

Micro measures act as tripwires to enable us to look at projects and processes and see if we can increase speed of actions and decrease time, cycle, and steps. While macro measures help us to focus on measuring the results of leadership on the corporate outcomes and to work the vision to see if the message is getting out there, micro measures help focus on the day-to-day routines and project activities.

Getting Started with a Logistics Council

The preliminary step to the creation of a logistics measurement system is the establishment of a Logistics Council or task force. This council is an intra-organizational team that is responsible for managing the supplier-customer relationships within the value chain. It consists of high-level management from each key organization. All Logistics Councils are charged with:

- Developing policy for the supply chain operations.
- Establishing an environment in which quality will be improved and sustained.
- Creating a quality mindset which focuses the organizations in the value chain enterprise on quality matters on a daily basis.

- Providing the leadership and personal involvement to guide Quality Management and Adaptive Engineering in its organizations.
- Developing its own Enterprise-wide Quality Improvement Plan that will enable it to carry out its specific charge.
- Review and evaluate the measurement system periodically and determine changes that may be needed.
- Perform management reviews of key activities performed by cross-functional teams within the value chain.

In summary, these guidelines are offered as useful tools and information that should be used where appropriate. They are not meant to be rules, rather ingredients from which you can pick and choose in order to supplement the things in which the management team is currently engaged. A Logistics Council provides the organizational power, wisdom and guidance to fuel an activity that has the potential for creating far-reaching changes in the way the entire enterprise operates.

Important Elements in Developing a Measurement System

Successful development of a measurement system requires fufillment of specific steps. Most of these steps will vary from organization to organization but the principal methods will be the same. It is useful at the beginning to have a corporate information system of some kind already in place in order to build on existing infrastructure and save development time. There are eight key steps involved in measurement system development:

1. Map out supply chain functions and objectives, products and services, customers and clients and their needs.

This activity involves the interaction of the strategy and operations of the supply chain's organizational units using the Process Management methodology previously described. The Process Management team is responsible for assuring that the functions of the operating area are in alignment with overall corporate strategy and direction. Enterprise-wide process mapping teams are an ideal way to map products, functions and processes to assure alignment and understanding.

When assessing the supply chain's products and services, it is often helpful to think of each member as an independent (but interconnected) organization that is in business to provide products and services. By asking and answering the following questions, a clear starting point for creating indicators can be obtained:

- Would we buy our services on the global market?
- How do we know?
- Can we compete as a global entrepreneur?
- Am I running my organization as a profitable and competitive logistical enterprise?
- What are my key indicators of satisfaction and cost of providing services?

It should be noted that in most cases, a common view of an organization's activities about internal relations does not exist. Since these activities have not been mapped, people have different opinions about the internal customer-supplier relationships that always exist. They are acting on the basis of different and incomplete maps of the reality which causes instant misunderstanding, lower productivity, and quality. Very few people ever have a clear picture of the totality. Business processes are seldom simple. In most cases, they consist of several processes that are connected into a network, requiring a common measurement system to provide the integration necessary for competitive advantage.

2. Determine appropriate outcome and process indicators for each major logistics function.

Generally speaking, there is at least one outcome indicator for each major logistics function. However, there can be many process indicators both within the function, as well as cross-functional. Generally speaking, there are seven major categories around which logistics indicators can be built:

- Community/client satisfaction
- Accuracy of services
- Timeliness
- Responsiveness or speed
- Cost of services or products

- Safety of the value chain environment
- Revenue or profitability of the value chain at large

Each major function should be analyzed to determine which outcome indices are appropriate. This will largely depend upon the purpose or objectives of the function being performed.

When identifying and building logistics indicators, it may be necessary to describe the process in terms of a flowchart, especially if one doesn't already exist. The goal is to develop a useful description of the process as it currently exists in order to identify key indicators and identify potential improvement areas.

When building indicators, it is important to look for interaction points among organizational units as it is at these interaction points that problems frequently occur. It is usually best to start with current measures of the logistics processes, where available. By envisioning what a successful supply chain looks like, missing indicators can be identified based upon the gaps and opportunities that exist. The key question to ask is, "How will the indicators help meet supplier/client needs and improve supply chain operations?"

3. Link all indicators back to the logistics objectives.

There are at least two ways to approach this activity: the short way and the linked way. The short way is to link all the indicators back to the individual corporate objectives. In many cases this means forcing a fit or showing no fit because of the many gaps that exist. The linked way is to first develop process level/activity level objectives which are linked to logistics objectives. The indicators are then linked to supply chain objectives, thus creating a tight fit. The process of linking indicators to objectives helps assure that the right things are being measured, and that they are being used to manage the community operations.

Templates, or flowcharts, are used to first link all existing measures to the corporate vision and objectives. Once existing indices are linked, then gaps and missing indices are identified and added to the system where appropriate. Decisions are also made on modifying or eliminating existing indices as new ones are being added. Overall, there are generally between 25 to 50 indicators that roll up into the "Balanced Scorecard."

4. Assign "indicator owners or champions."

In most cases, the "indicator owner/champion" should be the one person who either has the most ownership, has the most to lose, whose accountabilities are most affected by good performance, or all of the above. The point here is that the indicator owner has a major stake in the well-being of the "thing" that the indicator measures.

5. Focus on areas of logistics priority and create targets and goals.

When creating targets and goals in an objective manner, the following questions must be answered:

- What are we facing? (as in situation analysis)
- What do we as a Logistics Council want to happen? (objective setting)
- How are we going to get what we want? (alternative courses of action)
- What are the roadblocks and obstacles? (analysis of adverse consequences)

The targets and their relationships to objectives (whether corporate or community) should be apparent. Operating plans that are developed should also link to the target-goal whenever possible.

6. Focus teams to eliminate obstacles and drive data gathering.

The role of a logistics measurement team can be very useful in breaking down the initial barriers and resistance. Team members should be representative of the areas being measured and know where the data can be found. Getting reliable data to load into the logistics measurement system is often very time consuming and slow. Yet, the task of collecting meaningful data is essential to good measurement. It is easy to combine data that should not be mixed, and the possibilities for error are many.

This five-step data collection model is a good model to follow:

1. Clarify data collection goals
2. Develop operational definitions

3. Plan for data consistency and stability
4. Begin data collection
5. Continue improving measurement consistency and stability

Using measurement teams can often improve both the speed and accuracy, as well as give second-level ownership to the indicator data being assembled.

7. Load data and assess specific areas of concern.

A simple spreadsheet program can be used initially to load the historical data supporting the various community measures being tracked. Data should be entered into the system on a monthly basis, or more often, if possible. A comparison to targets can also be done monthly. (Hopefully, targets can be expressed monthly as well to facilitate this comparison.) Plans can then be developed to deal with longer-term problems.

8. Provide corrective action.

In an empowered, enterprise-wide logistical model, leadership involves group participation. Tight goals are sought by all levels and accepted both overtly and covertly. Strong pressure exists to get the facts and move swiftly to corrective action, while keeping everyone who has a need to know informed.

Summary

The following ten benefits can be accomplished by well-defined measures that are part of a Logistics Measurement system:

1. Identify the current logistics capabilities of the organization.
2. Highlight opportunities for process improvement and reengineering.
3. Facilitate goal-setting.
4. Help mark progress toward goal attainment.
5. Enable Benchmarking comparisons with other organizations.
6. Help improve job satisfaction and morale by enabling staff to work more effectively.
7. Place a strong emphasis on employee involvement.

8. Place emphasis on process, not people.
9. Produce higher quality products and enhanced pride in delivery.
10. Lower cost and increase productivity by harnessing the intelligence of everyone in the value chain.

Overall, the effective use of a well-designed Measurement system should yield a payback ratio of 4:1 or greater over a one year period following implementation. The Balanced Scorecard approach to measurement will help the Macrologistics strategy process to effectively integrate all key aspects of the supply chain. These measures provide the compass, dials and fuel gauge that lets us know what we are trying to achieve and how we are doing along the way. Logistics processes are a reality in any organization and we need to accept that fact. Then we can start to map the processes in order to identify them and understand them. That gives us the foundation for developing a measurement system for successful control and development of the Macrologistics Management process to achieve competitive advantage.

Mars Achieves Measurement Breakthrough in Strategic Alliance with CASS 11.1

The over-arching goal of the North American Logistics Services (NALS) division of Mars is to develop and deliver comprehensive logistics strategies that delivers the best possible level of service to our customers, at a price that is below competition. NALS helps to ensure responsive customer service by coordinating physical distribution from supplier to factory to warehouse to customer. Their slogan is: "We Transport."

NALS moves over eight billion pounds in a quarter of a million truckloads each year of raw materials, packaging, and finished goods throughout North America. Mars products occupy over five million square feet of storage in the U.S., Canada, and Mexico, as well as assisting operations in the Caribbean, Central and South America.

The NALS mission is to provide advanced logistics services and information, leading to improved customer satisfaction and supply chain integration such that a competitive advantage is created for Mars, Inc. A Strategic Alliance has been created with CASS to deliver logistics cost information back to the Mars business units in a timely and accurate manner. It links transportation and warehousing payment data with the customer and supplier data bases, as well as the order entry system, to provide a view across the pipeline and supply chain. The result is a logistics cost database that supports financial planning and reporting, tactical decision making, along with strategic initiatives such as network modeling.

Entering the 21st century, NALS is ready with technologies to provide real-time connectivity between suppliers, manufacturing logistics vendors, and customers. The Mars network of warehouses use bar coding scanning devices to speed inventory to customers. The Traffic Management system allows NALS to know the location and contents of all trucks transporting products. Also, warehouse managers know when products are scheduled to arrive from inbound trucks and coordinate a customer-bound trucks to avoid or minimize the cost and time-delay of warehousing.

On the measurement side of Macrologistics Management, a total of $1 million of logistics services are tracked each day, resulting in over 500,000 individual transactions per year. CASS audits, processes, pays, and reports on these transactions for Mars, acting as a third-party logistics agent. NALS on the other hand is responsible to collect and gather logistics information, analyze their cost and service drivers, and then take the proper action. Mars believes that real value is created when information contained in the order entry system is matched and integrated, to create a logistics cost database measurement system. This system is used to measure performance, identify variances, and take action. Ultimately, the transformation of data into information in a timely and accurate manner is the job of CASS Information Systems.

Over a six-year period, 3.2 million freight bills for $1.4 billion have been processed using an EDI match/pay system. Shipment and purchase order files reside on-line at CASS and serve as authorization flies for freight bill payment. Records in the authorization file are updated with payment information as bills are entered for payment and matched to the file. Each transaction is rated using the rates that NALS has negotiated with its vendors, and the rated amount is compared to the billed amount to detect any overcharges. All general ledger accounting is accommodated and NALS is provided with reports and accruals to assist them in the management of the Mars' logistics costs.

Critical Success Factors

The keys to Mars' success with the strategic alliance with CASS starts with communications. CASS has an in-plant rep and uses E-mail to provide a quick, paperless exchange of information. Also, access to senior management is readily available and involved at both Mars and CASS on a regular basis. The measurement system is used to report what is doing well, and not. Managing and fixing the exceptions is the key. Both organizations are on the lookout to reduce or eliminate paperwork and redundant activities. Next, expand the scope of services your vendor provides to you and that you provide your internal customers. What started with the payment of transportation invoices for Mars has expanded to include the payment and management reporting of warehousing export, raw materials/packaging and international transactions with multiple currencies, as well as other related tasks such as Benchmarking, general accounts payable, and certain administrative functions.

Next, impose discipline on your vendor and yourself. Handle information once and employ right first time measurements, concepts and thinking. Make sure that the data is accurate and consistent. Also, use EDI/electronic commerce to support Advanced Shipment Notifications (ANS) and Electronic Funds Transfer (EFT). Set a plan and timetable to get connected.

Look for ways to link suppliers, transportation, and warehousing vendors, as well as internal and external customers. Finally, remember mutuality. A mutual benefit is a shared benefit and will endure. Keep communication open, clear and consistent. Really learn from your alliance partner. Be sure to dedicate the necessary resources to foster the alliance, not only during the development and implementation stages, but into the future.

 Source: Council of Logistics Management, Annual Conference Proceedings, San Diego, California, October, 1995, pp. 335–349.

Air Products Uses Chemreg to Measure Compliance 11.2

With over 2,000 active products as well as several hundred new products each year, Air Products was very concerned about measuring its compliance to the U.S. DOT requirements surrounding the shipment and transportation of hazardous materials. The company had a manual process of generating shipping descriptions for bills of lading, labels, and materials safety data sheets. which was very labor intensive, time consuming, and error prone. Their business partners were getting less and less tolerant of 10 to 30 days required to establish a new product and constant scrutiny required an environment of continuous improvement in safety, health, and environmental performance.

In 1993, the company created a special logistics task team to create a system for measuring compliance to new regulations HM-181. This was the birth of CHEMREG, an internally developed software system having two separate but integrated applications:

- Structured product database containing physical characteristics on all of the commercial products and many experimental ones in the Chem Group, where data is inputted by various business areas as well as lab personnel
- Knowledge-based system that generates information, such as shipping descriptions and instructions, based upon product data and regulatory rules, in much the same way that an expert system would.

All appropriate regulatory information has been incorporated into the knowledge-based portion of CHEMREG, allowing it to produce accurate shipping descriptions for the following agencies: U.S. Department of Transportation (DOT), International Air Transport Association (IATA) and International Maritime Organization (IMO).

The CHEMREG team developed a new process for printing planning document instructions which assist the shipping locations in areas of proper labeling, placarding, and packaging instructions. A new global information system was added in 1994 to gather data more quickly and inform individuals of new shipping descriptions for newly established products. This reduced the cycle time from 7 to 30 days down to 2 days for the entire process. The measurement system also has a query capability that allows users to perform "what-if" analysis on product data and regulatory compliance.

Air Products has recently modified its knowledge-based expert system to accommodate another cycle of regulatory changes by DOT, IMO and IATA. This was followed by the automatic generation of Material Safety Data Sheets (MSDS). Using product data bases for product characteristics, an expert system determines appropriate phraseology and content for the MSDS. Eventually, CHEMREG is expected to become the focal point for distribution and Environmental, Health and Safety related regulatory issues around the world. Future topics to be addressed by the CHEMREG system include generating the content of the MSDS based upon European regulations, including ISO 14000 and related quality system standards.

 Source: Council of Logistics Management, Annual Conference Proceedings, San Diego, California, October, 1995, pp. 367–368.

Total Cost: A New Trend in Logistics 11.3

The call from purchasing managers and distributors is unanimous. The purchasing and distribution field has changed from one of inventory management and sales to a more holistic, value-added approach, featuring Macrologistic Management. The change that has had the most impact on the industry has been the shift toward a smaller supplier base, limiting the amount of vendors for each product and working more closely with suppliers. This new movement is called Total Cost, because it takes into account the total cost to the customer over and above the initial price. Late delivery, poor service, and faulty parts are looked on as part of the price of a product. Thus, industry leaders who have embraced Total Cost are willing to pay a higher initial purchase price when total cost benefits are guaranteed.

Therefore, the organizations on the brink of Total Cost breakthrough are selecting the "best and brightest" distributors. The primary benefits brought on by this change are improved efficiency and cost, more conclusive performance tracking, and stronger supplier relationships (or partnering). Distributors and purchasing professionals agree that many factors led to these changes, but the main catalyst has been an emphasis on value-added services. These services are generally a physical adaptation to a product such as an optional stereo for a car.

Whereas the most important demand that customers have for a product is for quality, their quality requirements vary sharply according to the customer's size, needs, and the extent to which the customers own quality system has developed. For example, Amoco demands quality from its distributors by monitoring distributor non-conformance through such variables as delivery, product quality, and administrative errors. Several priorities must be constantly looked at by the purchasing and distribution organizations for them to successfully embrace the changes around them.

Communication is a priority area where improvement is needed and new technological channels can help. Purchasers also note that distributors have fallen behind in education and technical ability. More specifically, purchasers want more new product information, better knowledge of the product, and most importantly, they want distributors to do their own screening to determine the best, most cost-efficient products.

Source: "Buyers and Distributors: It's Time for Teamwork," Christine Forbes, *Industrial Distribution(IND)*, August 1991, pp. 20–24.

12 Implementing Macrologistics Strategies: A Blueprint for Change

Transformation of businesses requires a sensitive understanding of an organization's vision, mission and values. All organizations undergo change. The use of specific strategies as a catalyst for change targeted to meet specific objectives is change management. Change management leadership is needed to identify the areas needing change and to provide direction to the change process. A structured change process utilizes specific strategies for each phase of the change process. For example, the alignment phase benefits from the use of a supply chain approach so that a clear blueprint for change can be designed.

The use of a Macrologistics strategy as the catalyst for change is valuable because it allows the change leader or "champion" to select an approach based on customer, supplier or logistics requirements. The Macrologistics strategy is non-threatening and does not require major changes in corporate culture. Further, there is little association with downsizing. In contrast, the strategy is designed to lead to new markets that should create not destroy job opportunities.

Before a change process can be implemented, it is important to consider the underlying values of the organization. In the case of General Electric, a shift to values that enhance stretch, speed and boundarylessness is considered desirable. In other organizations, the key values may be cost control or defect reduction. Synchronizing values among the members of the organization is a critical starting point. If the organization does not have common values it is difficult to embark on a change process.

The champion of the change process must start a process of dialogue that surfaces values and conforms them into an accepted set of values for the entire organization. Although many of the case studies described above included this value synchronization process often this was overlooked. Where it did occur, it was not documented well. The main point is that the change process should resolve value differences to go forward effectively into the alignment phase.

In some cases, maintaining the status quo is a value so that overcoming the organizational inertia implied by this value is a critical first step. Resistance to change is embedded in systems, business rules and data bases. Overcoming these elements of resistance is often crucial to the success of strategy implementation.

A Proposed Implementation Framework

Once values are clarified, the first phase of the change process begins. Figure 12.1 provides an overview of this process. As shown, the key phases of the change process are alignment, mobilization and integration. Also the process needs to have an evaluation component for monitoring progress and assessing results. As a model for organizational change, these phases represent an approach that is structured and that can be followed systematically to achieve change.

As indicated earlier, the capability to measure performance changes in areas related to the strategy is a critical requirement. For example, a strategy to reduce the time involved in order fulfillment can be measured by the relative amounts of time reduction achieved during each phase of implementation. Also, the goal can be presented in terms of reduction targets, e.g., 35% reduction in order-fulfillment times. (See Profile 8.4 for the Hewlett Packard case study.)

Each Macrologistics strategy has a change phase when it can be most effectively utilized. Figure 12.1 shows the strategies by phase. JIT II™ is effective during the alignment phase because it helps identify how suppliers can more effectively align with customer needs. The dialogue created internally with suppliers soon helps generate closer adherence to customer requirements throughout the organization.

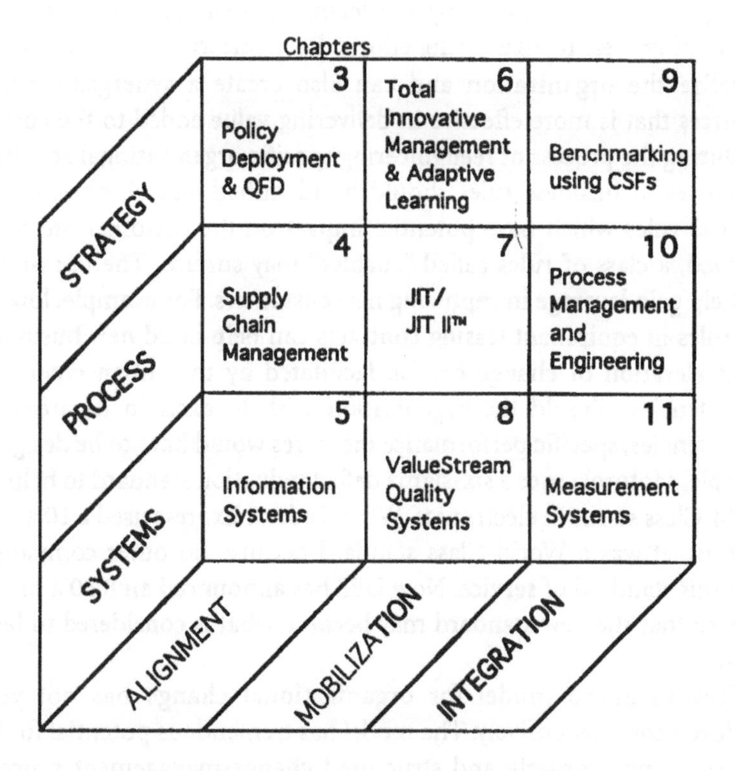

Figure 12.1 Macrologistics Management Alignment Matrix

Once there is alignment with customers and suppliers, the supply chain management approach is very useful because it can be used to provide a blueprint for how the customer needs are met throughout the organization. This blueprint can build on the information developed in the JIT II™ process. When the blueprint is being prepared, Benchmarking can be used with other companies to make certain that the blueprint is correctly defined. The use of an outside perspective helps to gain insights on how companies differ in their approaches to meeting customer needs.

Mobilization is a part of the change process which often is dealt with on a budgetary basis. Change requires more than the budget. Teams of knowledgeable employees have to be organized to target the use of resources. Here the use of strategic partnerships and providers of third-party logistics support can be of great help. Their expertise and knowledge can stretch the use of company resources and free up resources for internal improvements.

Gaps that are identified by the teams in their discussions become the subject of process reengineering efforts. Reengineering these processes helps revitalize the organization and can also create a synergistic change in resources that is more effective in delivering value added to the customer.

During the process of reengineering specific organizational activities, discrepancies in business rules should be identified and eliminated. If those business rules which have potential impact on the customer are separately analyzed, a class of rules called "enabler" may surface. The use of enablers can help gain leverage in capturing new customers. For example, low interest rate rules in equipment leasing contracts can help build new business.

Acceleration of change can be facilitated by the enhancement of core competencies. Should the organization wish to excel in its areas of core competencies, specific performance measures would have to be designed. For example, Motorola uses a six-sigma defect reduction standard to help achieve World-Class status in electronics. When Federal Express used a 10 a.m. delivery time, it was a World-Class standard because no other company could meet this standard of service. Now UPS has announced an 8:30 a.m. delivery time, so that the new standard may become what is considered to be World Class.

This advanced model for organizational change has not yet been employed comprehensively. The model has tremendous potential for helping develop a more orderly and structured change management process. The above case studies show how each strategy can be used separately and it is anticipated that as the high cost of disorganized change becomes apparent, more companies will desire to control change and plan for these strategies to be used when they can be most effective.

One barrier to the implementation of this approach is the high cost of overhauling the information system. Often, the information systems department is the last to have voice in the changes which are planned. If the change process is timed to coincide with information systems overhaul, then new information requirements can be more readily accommodated. These information system requirements optimally would include continuous performance measures, built in evaluation for each strategy and feedback that is actionable and delivered promptly to the organizational elements responsible for these actions.

One approach for design of the overall change process is defined as "adaptive engineering". The concept borrows the term adaptive from Peter Senge at MIT who uses this concept to show a continuous learning approach to problem solving. It combines this term with reengineering approaches pioneered by Michael Hammer at Harvard. What is needed is a continuous tool for assessing and refining change strategies without the need to stop ongoing processes. The need for a continuous process for quality improvements is critical to the success of the change process. TQM, therefore, often uses a Plan-Do-Check-Act cycle of decisions to implement quality improvements.

Adaptive engineering uses the sequence Decide-Test-Evaluate-Change. This approach permits the systematic use and evaluation of Macrologistics strategies. Change management is facilitated and there is ample opportunity to learn from the applications of the strategy what is working and what strategy needs to be revised or even discarded.

Developing World-Class Macrologistics Goals

The use of these systematic approaches to change management or adaptive engineering has an additional benefit. The result of consistent application of these strategies can help an organization define what it will take to be a World-Class competitor. This happens because Macrologistics strategies naturally evolve into stretch and speed types of goals. If these goals are prioritized, the organization can generate substantial competitive difference in key deliverables to customers. If service or product is delivered that is far in excess of what competition is able to provide, there is a definite opportunity for competitive advantage to occur.

Several strategies are likely to improve the chance to make this happen. In the alignment phase, Benchmarking needs to be performed in a way to identify the crucial element of World-Class competitors. Alignment can also be used to verify that these elements are valuable to customers and that suppliers and employees agree about the importance of these elements. For example, if order fulfillment is critical, then a major change in order fulfillment is likely to surpass competition and vault companies like Hewlett Packard into World Class status. Recently, *The Wall Street Journal* hailed both Hewlett Packard and Motorola as examples of American World-Class Competitors.

Even what appears to be a minor change in logistics policy can have a major impact on market expansion opportunities. *The Wall Street Journal* cites both Compaq and WalMart as companies who used the warehousing and delivery capability of suppliers to help generate cost and time savings. For Compaq improvements to delivery times by a supplier, Phelps Tool & Die Co. are cited as major cost reduction factors that help Compaq produce the new ProLinea computer less expensively in Houston than in Taiwan Competitors of Compaq are said to fear these cost reduction factors so that Apple and IBM may see loss of business to Compaq as a result. (June 17, 1994).

The Macrologistics strategy approach offers substantial opportunities to bring the voice of the customer, supplier and the employee into alignment. This is what happens during the JIT II™ process and when the supply chain is blueprinted. Mobilization strategies often involve team work that helps empower change and the use of business rules enablers helps create leverage factors to accelerate the change process. Using the adaptive engineering approach as an overall change management tool means that those strategies that deliver substantial improvements on the critical areas of core competency and core output are continuously refined and improved.

Macrologistics strategy is itself an area of core competency that needs enhancement. How can managers develop stretch, speed and boundaryless-ness types of goals without it? Yet, it appears that not many top executives and board members value the improved potential provided by these valuable strategic approaches to the change process.

The purpose of this book has been to present the existing case studies in this new and emerging area of logistics management in an organized fashion. Also, we suggest ways to orchestrate their application. By systematically employing these strategies, it is possible to develop more effective future applications.

Now that you are the CEO or at least you have to think like one, the change process and the method for selecting change catalysts becomes easier. We believe that these tools for change and the systematic approach for their application in large organization will help companies develop new and breakthrough approaches. They can use these approaches for simple and quick solutions to complex logistics problems or they can use them as a framework for change. Becoming a World-Class competitor means development of systematic and scientifically employed strategic change processes. We feel that the evidence is very strong that the Macrologistics strategy offers a unique opportunity to change in a way that delivers World-Class competitive status.

The Action Plan

- Define the objective of the strategy, why it is important, and how its effectiveness will be measured.

- Define the information system strategy changes necessary to implement the strategy.
- Use the project management format of the 5 W's and the 1 H to develop the plans.
- Create adaptive learning processes to continuously improve the understanding of the problem.
- Create adaptive engineering processes to continuously refine the solutions.
- Continue to assess core competencies, core processes, and core outputs.

Development of a macrologistics strategy plan is something that will be unique to each organization. The strategies selected will depend on the current state of the change process and the degree to which the organization desires to achieve World-Class status in its area of core competency and core output.

The kinds of strategies that have been described in the case studies demonstrate how individual strategies have been used by companies such as Xerox to generate a billion dollars of bottom line profits or by Hewlett Packard to become a world-class competitor.

The innovative Macrologistics approach to change management can help you select catalysts for change that are effective, non-threatening and that have the potential for radically enhancing your company's capability to achieve world class status. This is the kind of breakthrough change that will be needed as we enter the 21st century change that is both innovative and implementable.

Figure 12.2 depicts the fourteen steps involved in a fully integrated Macrologistics Management. system. The timing and introduction of each step in many cases depends on the state of logistics development within each organization and the relationships shown between steps are based on a typical implementation journey of three years.

The challenge for management is to begin the process by using an adaptive engineering approach for selecting new strategies. Once the strategies are tested and evaluated, they can be continuously refined. The process of managing change can be systematic and can be designed to provide organizational speed and stretch breakthroughs. We believe that these strategies offer you and your company an opportunity to achieve breakthrough change that can and will make a difference. Although this process is not a guaranteed ticket to the CEO's job, these concepts are powerful enough to help turbocharge the logistics journey.

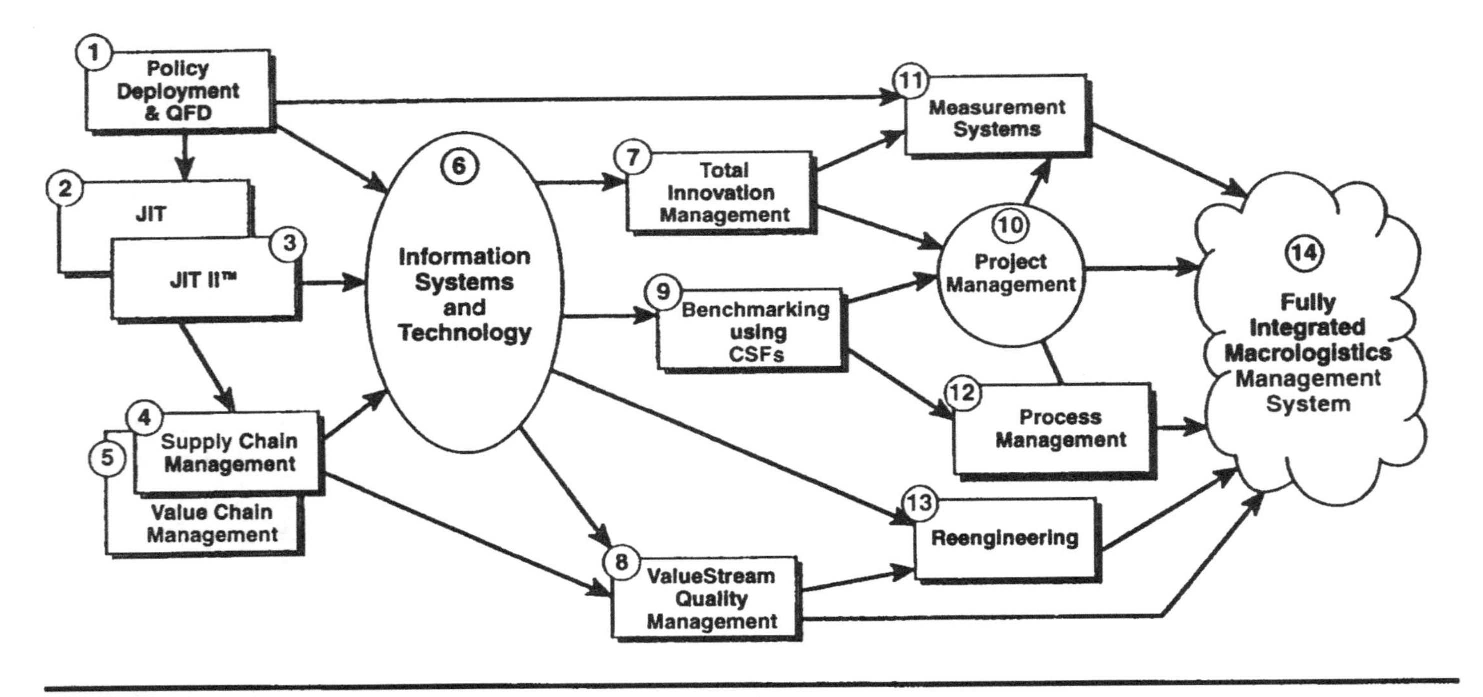

Figure 12.2 Macrologistics Management Implementation Fourteen-Step Integration Diagram

Glossary*

360 Degree Review: Performance review that includes feedback from superiors, peers, subordinates, and clients.

Abnormal Variation: Changes in process performance that cannot be accounted for by typical day-to-day variation. Also referred to as nonrandom variation.

Acceptable Quality Level (AQL): The minimum number of parts that must comply with quality standards, usually stated as a percentage.

Activity: The tasks performed to change inputs into outputs.

Adaptable: An adaptable process is designed to maintain effectiveness and efficiency as requirements change. The process is deemed adaptable when there is agreement among suppliers, owners, and customers that the process will meet requirements throughout the strategic period.

Adaptive Engineering: An evolutionary approach to process reengineering that uses continuous improvement techniques to identify innovative solutions to problems. The solutions are adapted for maximum leverage and are designed to become catalysts for change.

Appraisal Cost: The cost incurred to determine defects.

Annual Quality Report: The annual quality report focuses on quality performance. The traditional annual report focuses on functional performance. They can be separated or integrated documents.

Approach: A Baldrige Award criteria of the way quality is designed into a process.

ASQC: The American Society for Quality Control, Quality Assurance and Total Quality Management, consisting of 150,000 world-wide members. Publishes *Quality Progress* magazine.

---

* This glossary, developed by Strategy Associates in 1993 and updated in 1997, has been generated to describe almost 150 integration points possible between Total Quality and Macrologistics Management. (Use the index for more detailed macrologisics references.)

Baldrige Award: The Malcomb Baldrige National Quality Award is a highly competitive, prestigious yearly award given to a maximum of two service, two manufacturing, and two small businesses that have demonstrated excellence in meeting the requirements of the rigorous award criteria administered by the U.S. Department of Commerce, National Institute of Standards and Technology (NIST).

Benchmarking: A tool used to improve products, services, or management processes by analyzing the best practices of other companies to determine standards of performance and how to achieve them in order to increase customer satisfaction.

Business Objectives: Specific objectives that, if achieved, will ensure that the operating objectives of the organization are in alignment with the vision, values, and strategic direction. They are generally high level and timeless.

Business Process: Organization of people, equipment, energy procedures, and material into measurable, value-added activities needed to produce a specified end result.

Business Process Analysis (BPA): Review and documentation (mapping) of a key business process to understand how it currently functions and to establish measures.

Business Rules: An advanced concept in information system design applied to Macrologistics to help structure the information system around strategic rules. These business rules are enablers and inhibitors and when they are isolated can provide valuable leverage for change.

Champion: A manager who oversees specific quality improvement projects, and assists them in obtaining proper resources and buy-in. Same as **Sponsor.**

Competitive: A process is considered to be competitive when its overall performance is judged to be as good as that of comparable processes. Competitiveness is based on a set of performance characteristics (defects, costs, inventory turnaround, etc.) that are monitored and tracked against comparable processes within the corporation, the industry, and/or the general business community.

Competitive Benchmarking: Comparing and rating the practices, processes, and products of an organization against the world best, best-in-class, or the competition. Comparisons are not confined to the same industry.

Conformance: Affirmative indication or judgment that a product or service has met specified requirements, contracts, or regulations. The state of meeting the requirements.

Complaint Tracking: Detailing when complaints come in, what is done about them, and when they are closed. Several software packages can be used to assist in this process.

Continuous Improvement: Sometimes called Constancy of Purpose, this is a principle used by W. Edwards Deming to examine improvement of product and service. It involves searching unceasingly for ever-higher levels of quality by isolating sources of defects. It is called *kaizen* in Japan, where the goal is zero defects. Quality management and improvement is a never-ending activity.

Control: The state of stability, or normal variation and predictability. It is the process of regulating and guiding operations and processes using quantitative data. Control mechanisms are also used to detect and avoid potential adverse effects of change.

Control Charts: Statistical plots derived from measuring a process. They help detect and determine deviations before a defect results. Inherent variations in manufacturing processes can be spotted and accounted for by designers.

Corrective Action: The implementation of effective solutions that result in the elimination of identified product, service, and process problems.

Cost-of-Quality: The sum of prevention, appraisal, and failure costs, usually expressed as a percentage of total cost or revenue.

Critical Success Factors (CSFs): Areas in which results, if satisfactory, will ensure successful corporate performance. They ensure that the company will meet its business objectives. CSFs are focused, fluctuate, and are conducive to short-term plans.

Cross-Functional: A term used to describe individuals from different business units or functions who are part of a team to solve problems, plan, and develop solutions for process-related actions affecting the organization as a system.

Cross-Functional Focus: The effort to define the flow of work products in a business process as determined by their sequence of activities, rather than by functional or organizational boundaries.

Culture: Values that permeate the organization. A culture is communicated by hero stories, by the reasons people get promotions and recognition, by hall-talk, and the questions that are asked by the Leadership Team. A Service Quality culture is one that is rigorous, customer focused, and values employees.

Customer: The recipient or beneficiary of the outputs of work efforts or the purchaser of products and services. May be either internal or external to the company.

Customer, Internal: Organizations have both external and internal customers. Many functions and activities are not directly involved with external customer satisfaction, but their outputs provide inputs to other functions and activities within the organization. Data processing, for example, must provide an acceptable quality level for many internal customers.

Customer Value: A term coined by Bradley Gale in Managing Customer Value that relates the customer's perception of the services relative to the competition and price paid for these services.

Customer Requirements (also called valid requirements): The statement of needs or expectations that a product or service must satisfy. Requirements must be specific, measurable, negotiated, agreed to, documented, and communicated.

Customer/Supplier Model: The model is generally represented using three inter-connected triangles to depict inputs flowing into a work process that, in turn, adds value and produces outputs that are delivered to a customer. Throughout the process, requirements and feedback are fed from the customer to the supplier to ensure that customer quality requirements are met.

Cycle Time: The elapsed time between the commencement and completion of a task. In manufacturing, it is calculated as the number of units of work-in-process inventory divided by the number of units processed in a specific period. In order processing it can be the time between receipt and delivery of an order. Overall cycle time can mean the time from concept of a new product or service until it is brought to market.

Defect: Something that does not conform to requirements.

Deployment: A term frequently used in the Baldridge Award criteria to mean how thoroughly and how well the quality processes permeate the organization. Consistency is one of the keys.

Document of Understanding (DOU): A formal agreement defining the roles, responsibilities, and objectives of all the parties to that agreement. The degree of detail is dictated by the nature of the agreement, but it should always clearly address the requirements of the work product in question.

Effective: An effective process produces output that conforms to customer requirements. The lack of process effectiveness is measured by the degree to which the process output does not conform to customer requirements (that is, by the level of defect of the output).

Effectiveness: The state of having produced a decided or desired effect; the state of achieving customer satisfaction.

Efficiency: A measure of performance that compares output production with cost or resource utilization (as in number of units per employee per hour or per dollar).

Efficient: An efficient process produces the required output at the lowest possible (minimum) cost. That is, the process avoids waste or loss of resources in producing the required output. Process efficiency is measured by the ratio of required output to the cost of producing that output. This cost is expressed in units of applied resource (dollars, hours, energy, etc.).

Employee Involvement (EI): Promotion and mechanisms to achieve employee contributions, individually and in groups, to quality and cost reduction performance objectives. Cross-functional teams, task forces, quality circles, or other vehicles for involvement are used.

Employee Well-Being and Morale: Maintenance of work environment conducive to well-being and growth of all employees. Factors include health, safety, satisfaction, work environment, training, and special services such as counseling assistance, recreation, etc.

Executive Quality Service Council (EQSC): Comprised of members of executive management and union leadership who oversee the quality effort from a corporate view and set strategic direction.

Facilitator: Responsible for guiding the team through analysis of the process. Also concerned with how well the team works together.

Failure Cost: Cost resulting from occurrence of defects (such as scrap, rework/redo, replacement, etc.).

Functional Organization: An organization responsible for one of the major corporate business functions such as marketing, sales, design, manufacturing, or distribution.

Gap Analysis: Comparing hard reality against goals.

Human Resource Management: Development of plans and practices that realize the full potential of the workforce to pursue the quality and performance objectives of the organization. Includes (1) education and training, (2) recruitment, (3) involvement, (4) empowerment, and (5) recognition.

Implementer: An individual working within the process and who is responsible for carrying out specific job tasks.

Indicators: Benchmarks, targets, standards, or other measures used to evaluate how well quality values and programs are integrated.

Industry Trend Analysis: Trends that are taking place in the whole industry. This is important in service because the bar keeps rising on customer expectations, satisfaction and needs.

Information System: A database of information used for planning day-to-day management and control of quality. Types of data should include (1) customer related, (2) internal operations, (3) company performance, and (4) cost and financial.

Inputs: Products or services obtained from others (suppliers) in order to perform job tasks. Material or information required to complete the activities necessary for a specified end result.

Involved Managers: Managers who have responsibility for the day-to-day activities and tasks within the process.

ISO: A worldwide federation of national standards bodies representing almost 100 countries. It promotes the development of standardization and related activities to facilitate the international exchange of goods and services. It consists of 173 technical committees, 1,830 sub-committees, and 18 ad hoc groups.

ISO 9000 Standards: A series of standards generated by the International Organization for Standardization. All told, there are over 8,000 standards published by ISO, representing 70,000 pages of technical text.

Just-in-Time (JIT): The delivery of parts and materials by a supplier at the moment a factory needs them, thus eliminating costly inventories.

JIT II™: A major advance over JIT because it creates institutions for implementing JIT. The process uses supplier partnering and concurrent engineering to develop new opportunities for business leverage and it also supports mobilization. The tangible elements are an inplant, an evergreen supplier contract, and integrated communications systems. Intangibles include improved cooperation toward developing competitive advantage and enhancement of core competencies. The JIT II™ process also includes licensing opportunities so that suppliers can create a keiretsu-like network with other potential business partners.

Joint Planning: A planning process that includes the enterprise, its suppliers, and its customers.

Kaizen: A Japanese management term meaning continuous improvement.

Key Processes: The most critical processes to customer satisfaction and the survival of the enterprise. Also known as **Core Processes.**

Key Quality Indicators: Those areas that customers have indicated are vital to their satisfaction.

Key Quality Objectives: Goals that pertain to key quality indicators.

Leadership: The category of the Baldridge Award that examines personal leadership and involvement of executives in creating and sustaining customer focus and clear and visible quality values.

Macrologistics: A new concept pioneered by Dr. Martin Stein to help differentiate from the tactical application of logistics. The Macrologistics strategy targets the use of logistics changes for strategic purposes and help organizations obtain competitive advantage. The process is used to define and create breakthrough results and develop World -Class competitors.

Management Cycle: Consists of Vision, Strategy Planning, Organizing, Implementing, and Controlling. It is recognized as a key component of Management Quality.

Management for Quality: The translation of customer focus and quality values into implementation plans for all levels of management and supervision.

Measurable Outcomes: Specific results that determine on an organizational basis how well the critical success factors and business objectives are being achieved.

Measurement: The methods and system of indicators used to achieve and maintain conformance to customer requirements. Examples of corporate systems are the Balanced Scorecard, House of Quality System of Indicators, and Critical Success Factor Measures.

Mission: The core purpose of being for an organization. Usually expressed in the form of a statement twenty-five to fifty words in length.

Operating Plans: Specific, actionable plans which, if carried out successfully, ensure that critical success factors are met. This will lead to corporate business objectives being met. They are tied to the Critical Success Factors, are detailed, and contain measurements of success.

Operating Quality Service Council: Comprised of regional or division executives and their direct reports, and may include union and staff representation. The council oversees the quality effort within the organization and ensures that quality strategies support the organization's strategic direction.

Operational Results: Results that relate to the operation of the enterprise. Those results may be financial or quality measures.

Paradigm Shift: A fundamental change in the way one views the world.

Pareto Principle: Coined by Dr. Joseph Juran. In any phenomenon, only a few of the contributors account for the bulk of the effect. Also known as the 80-20 rule governing the trivial many and the vital few.

Perceptual Measure of Satisfaction: Surveys that measure a customer's perception of the service or product. These measures may include surveys, focus groups, and direct observations.

Policy: A guide to management and employees action.

Policy Deployment: Is a planning system adapted from the Japanese Hoshin Kanri, meaning the alignment of the shiny goals or arrows. Is a key component of Logistics Management in many service organizations.

Prevention Activity: Elements of prevention activity include (1) education in process quality management and (2) process management (ownership, documentation/analysis, requirements activity, measurements including statistical techniques, and corrective action on the process).

Prevention Cost: Costs incurred to reduce the total cost of quality.

Process: The organization of people, equipment, energy, procedures, and material into the work activities needed to produce a specified end result (work product). A sequence of repeatable activities characterized as having measurable inputs, value-added activities, and measurable outputs. It is a set of interrelated work activities characterized by a set of specific inputs and value-added tasks that produce a set of specific outputs.

Process Analysis: The systematic examination of a process model to establish a comprehensive understanding of the process itself. The intent of the examination should include consideration of simplification, elimination of unneeded or redundant elements, and improvement.

Process Control: The activity necessary to ensure that the process is performing as designed. Achieved through the use of statistical techniques, such as control charts, so that appropriate actions can be taken to achieve and maintain a state of statistical control.

Process Elements:A process is comprised of activities and tasks. A process may also be referred to as a subprocess when it is subordinate to, but part of, a larger process. A subprocess can also be defined as a group of activities within a process that comprise a definable component.

Process Management: The disciplined management approach of applying prevention methodologies to the implementation, improvement, and change of work processes to achieve effectiveness, efficiency, and adaptability. Critical to the success of process management is the concept of cross-functional focus.

Process Model: A detailed representation of the process (graphic, textual, mathematical) as it currently exists.

Process Owner: Coordinates the various functions and work activities at all levels of a process, has the authority or ability to make changes in the process as required, and manages the process end-to-end so as to ensure optimal overall performance.

Process Performance Quality: A measure of how effectively and efficiently a process satisfies customer requirements. The ability of a product or service to meet and exceed the expectations of customers.

Process Review: An objective assessment of how well the methodology has been applied to the process. Emphasizes the potential for long-term process results rather than the actual results achieved.

Quality Accountability System: A methodology that incorporates quality goals into the practices of hiring, promotion, performance appraisals, and compensation.

Quality Assessment System: The system of assessing the service delivery processes in order to improve that process at regular intervals.

Quality Assurance: An independent evaluation of quality-related performance, usually done for those who have a need to know rather than for those directly involved.

Quality Control: A process that evaluates the actual performance against goals and that takes action based upon the difference.

Quality Function Deployment (QFD): A system that pays special attention to customer needs and integrates them into the marketing, design, manufacturing, and service processes. Activities that do not contribute to customer needs are considered wasteful.

Quality Improvement Team (QIT): A group of people brought together to resolve a specific problem or issue identified by a business process analysis, individual employees, or the Operating Quality Service Council. A group of individuals charged with the task of planning and implementing process quality improvement. The three major roles in this task force are team leader, team facilitator, and team member.

Quality Journey: A ten-step structured problem-solving model which is used to deploy success stories to management, employees, suppliers, customers, shareholders, and the community. It is an offshoot of the QC Story, originated by Dr. Kume of JUSE.

Quality Management: The management of a process to maximize customer satisfaction at the lowest overall cost to the company.

Quality Management System: The collective plans, activities, and events established to ensure that a product, process, or service will satisfy given needs. The infrastructure supporting the operational process management and improvement methodology.

Quality Planning: The process of developing the quality master to link together all of the planning systems of the organization. The objective is to follow all areas of achievement of the vision, mission, and business objectives and to operationalize the strategy by identifying the requirements to achieve leadership in the market segments chosen. Includes key requirements and performance indicators and the resources committed for these requirements.

Quality Readiness: Describes the cultural readiness of an organization to embrace and implement a Service Quality System.

Quality Tool: Instrument or technique that supports the activities of process quality management and improvement.

Reengineering: Is an advanced form of Quality Management that designs the organization from the ground zero level with the customers' preferences foremost in mind.

Requirements: What is expected in providing a product or service. The *it* in "do it right the first time." Specific and measurable customer needs with an associated performance standard.

Resource Allocation: A decision to allocate resources, capital, and people to support specific operating plans, tied to the budget process.

Results: Results are, quite simply, a measurement of how well corporate business objectives are being met. Results require that standards and goals for performance are set and the results or processes and performance tracked.

Robust Design: Making product designs "production-proof" by building in tolerances for manufacturing variables that are known to be unavoidable.

Root Cause: Original reason for nonconformance within a process. When the root cause is removed or corrected, the nonconformance will be eliminated.

Service: Work performed for others. Support services within an organization, such as payroll, engineering, maintenance, hiring, training, etc.

Service Design: Consciously designing the service to meet the needs of the customers at the lowest cost.

Six Sigma: A statistical term that indicates a defect level. One-sigma means 68% of products are acceptable; three-sigma means 99.75; and six-sigma means 99.999997% perfect or 3.4 defects per million parts.

Sponsor: Advocate for the team who provides resources and helps define mission and scope to set limits.

Stakeholder: Individual or department who either has an effect on the process or is affected by it.

State of the Organization: A report that gives a snapshot of the organization in its current state. Usually it contains a section on the financial, quality, human resource, community, and competitive status. It is used to do a gap analysis so that the strategic plan is a basis in fact from which to operate.

Statistical Process Control (SPC): The use of statistical techniques, such as control charts, to analyze a work process or its outputs. The data can be used to identify deviations so that appropriate action can be taken to maintain a state of statistical control (predetermined upper and lower limits) and to improve the capability of the process.

Statistical Quality Control (SQC): A method of analyzing measured deviations in manufactured materials, parts, and products.

Strategic Quality Planning: Development of strategic and operational plans that incorporate quality as product or service differentiation and the load bearing structure of the planning process. Includes (1) definition of customer requirements, (2) projections of the industry and competitive environment for identification of opportunities and risks, and (3) comparison of opportunities and risks against company resources and capabilities.

Sub-processes: The internal processes that make up a process.

Suggestion System: A systematic way to solicit ideas for improvement from employees. The means can be a passive box on wall or active meetings, e-mail. Prizes are usually given for bright ideas implemented.

Suppliers: Individuals or groups who provide input. Suppliers can be internal or external to a company, group, or organization.

Supplier Quality Management: Managing the quality of the services or products of suppliers. Most high-performing SQS organizations both extend their SQS training to suppliers and have rigorous measurement systems in place to assess their performance.

Supply Chain Management: The systematic effort to provide integrated management to the Supply Value Chain in order to meet customer needs and expectations, from suppliers of raw materials through manufacturing and on to end-customers.

System: A set of processes with a purpose.

Systems Thinking: A school of thought that uses feedback and computer simulation software for training and strategy support. **Strategy System** software is a culmination of a future vision where systems, tools and techniques are used to peer onto the future and look at the implications of the assumptions behind our mental models. The new field was popularized by Peter Senge in his book, ***The Fifth Discipline.***

Taguchi Methods: Statistical techniques developed by Genichi Taguchi, a Japanese consultant, for optimizing design and production.

Task: The basic work element of a process activity.

Top-Down Improvement: Improvements that emanate from the Policy Deployment or Strategic Planning process. Ideally, each customers needs and satisfaction is integrated into that process. Each department, unit, or crossfunctional team then sets goals that will help achieve the overall key quality objective.

Total Innovative Management: A system of Management Quality containing 12 factors involved in excellent corporate performance, of which the management cycle is the most critical.

Total Quality Management (TQM): The application of quality principles for the integration of all functions and processes of the organization. The ultimate goal is customer satisfaction. The way to achieve it is through continuous improvement. Also known as Company-Wide Total Quality Control. Usually consists of Teams, Policy Deployment, and Daily Control.

Value-Added Assessment: A sorting-out process used to distinguish between real business value-added activities vs. those that are non-value-added. It is considered an essential tool for improving the effectiveness and efficiency of business processes.

Value Chain: The extended enterprise of an organization, including suppliers and distributors, that constitute the chain of value-added services offered to customers and stakeholders.

ValueStream Quality: A Quality System developed by Frank Voehl and Jeffrey Vengrow to describe downment and improve the quality system of an organization's extended enterprise.

Variation: The degree to which a product, service, or element deviates from the specification or requirements. Quality in service organizations deals with identifying, measuring, and adjusting to variability resulting from interactions with customers, while manufacturing organizations are focused on bringing product variability under control.

Visibility Plan: A conscious effort to make improvement results visible to all employees, suppliers, stakeholders, and customers. Usually includes how to communicate success stories via newsletters, press releases, meetings and e-mail.

Vision: The long-term future desired state of an organization, usually expressed in a 7- to 20-year time frame. Often included in the vision statement are the areas that the organization needs to care about in order to succeed. The vision should inspire and motivate.

INDEX

Index

A

Account leader, ARCO project, 126
Acquisition costs, Becton Dickinson, 71
Action plan, macrologistics strategies implementation, 251
Action verbs, 127–30
Activity-based accounting, 6
  total innovative management, 119
Adams, W.A., 221
Adaptive engineering, 23, 201, 234, 248, 249, 250, 251
  macrologistics strategy, 11
  total innovative management, 118
Adaptive learning, 23, 159, 251
  Burlington Northern, 121
  management quality, innovation, integration of. *See* Total innovative management
  total innovative management, 117
Added value, 182
  return-on-quality software system, 191
Advanced logistics services, 240
Advanced shipment notifications, 241
Aggregate production planning, 163
Air Products, balanced scorecard, 242–43
Akao, Yoji, 29
Alignment, 11, 13, 16, 17–18, 19, 20, 21, 149, 197, 199, 245, 246, 247, 249
  of customer needs, 33
  information technology, 81
  supply chain management and, 61, 62
  Taguchi methods and, 37
  total innovative management, 118
Alliances, 181
  strategic, total innovative management, 114
American Productivity and Quality Center, 212
ANS. *See* Advanced Shipment Notifications
Applied research, Eastman Chemical, 135
APQC. *See* American Productivity and Quality Center
Architecture, 98–99
ARCO project, 125
Artificial intelligence, 211
Asset-change management, 211
AT&T, critical success factors, 200–202
Automated sales, 211
Automatic replenishment system, Becton Dickinson, 73
Automation, 159, 172, 176
Autonomy, 229

B

Back order, 220
Background processes, 219
Balanced scorecard, corporate measurement system, 231–44
  Air Products, 242–43
  Chemreg, 242–43
  logistics council, 234–35
  Mars, 240–42

measurement system, developing, 235–39
measures, types of, 234
total cost, 243–44
Balanced scorecard measurement system, 22, 231, 234, 237, 239
Baldridge, Malcolm, 19. *See also* Malcolm Baldridge National Quality Award
Baldrige Award, 99, 225
Bar coding, 151, 240
Battaglia, Alfred J., 60
Becton Dickinson, 69–73, 72
background information, 69–70
change, conditions for, 70–72
quality, 69
supply chain management, 70
vision, 69
Bell Labs, just-in-time management, 169–70
Benchmark Clearinghouse, 212
Benchmarking
Becton Dickinson, 73
defined, 204–5
logistics, steps to, 206–7
partners, 212
team, 206
types of, 205
Bennis, Warren, 56, 57
Best practices, 208, 212
room, Rubbermaid, 132
Bose, just-in-time management, 160–67
Boundarylessness, concept of, in macrologistics strategy, 5, 8, 245, 250
Bowersox, Donald J., 63
Brainstorming, 138, 146, 227
in quality function deployment, 30
Breakthrough approach, 29, 121, 129, 130, 137, 153, 167, 168, 195, 198, 201, 215, 216, 217, 218, 221, 233, 250, 251
ARCO project, 126
Burlington Northern, 121, 123
FPL Group, 43
in macrologistics strategy, 22, 27
supply chain management and, 57, 59
Breakthrough goal-setting, 128
Breakthrough learning process, 130
Breakthrough team, 129
ARCO project, 125, 126
Burlington Northern, 121
Moorhead Malting Facility, 124
Bulletin boards, 90–91
Burlington Northern, 121–30
ARCO project, 125–27
background information, 121–22
Moorhead malting facility, 124–25
small-scale teams, 122–23
urgent measurable goals, 123
Business centers, Eastman Chemical, 135
Business planning, 163
software, total innovative management, 119
Business process, 216, 236
innovators, 212
management, 216, 217, 218
lifecycle, 218
redesign, 218, 219
reengineering, 96, 217–18
Business rules, 248
synergy and, 8
Business rules analysis, 19, 23
in macrologistics strategy, 18
Business rules enabler, 250
Business structure, total innovative management, 114
Business transformation
Becton Dickinson, 72
total innovative management, 109

C

Capability studies, 156
Case studies, Taguchi, 35
Catalyst, 164, 245
for change, 153
Catchball
FPL Group, 44
in policy deployment, 28
in quality function deployment, 31
Cause-and-effect analysis, 38
principles, in quality function deployment, 30
Certified vendor, 158
Champion of change process, 246
Champy, James, 218
Change
accelerator, total innovative management, 119
catalyst for, in macrologistics strategy, 12

catalysts, 250
leader, 245
Change management, 212, 245, 249, 250, 251
Becton Dickinson, 70, 72
model, 16–17
Becton Dickinson, 72
Change process, 245, 249, 250
Channels, total innovative management, 114
Charter boundaries, ARCO project, 126
Chemreg, balanced scorecard, 242–43
Chrysler Corporation
chimneys, 143–44
culture, changes in, 144–45
leadership style, changes in, 144–45
macro structure, changes in, 144–45
management systems, changes in, 144–45
platform teams, 143–44
CMS. *See* Corporate measurement system
Collaborative networks, 91–92
Command theater, 211
Commercial Darwinism, 175
Commercial dominance, 175
Commercialization, Eastman Chemical, 134
Community/client satisfaction, 236
Competition, 138, 143, 186
Becton Dickinson, 72
Competitive advantage, 143, 160, 167, 175, 176, 177, 182, 183, 184, 188, 191, 195, 198, 203, 204, 205, 219, 236, 240
3-M, 51
Becton Dickinson, 69
supply chain management and, 60
total innovative management, 118, 119
Competitive benchmarking, 205
ARCO project, 125
Competitive challenge, 217
Competitive difference, 249
Competitive environment, supply chain management and, 59
Competitive intelligence, 197, 199
Competitive logistical enterprise, 236
Competitive market, 192
Competitive marketing analysis, critical success factors, 197–98
Competitive model, 180
Competitive performance, 220
Competitive positioning, 200, 218
Competitive strategy, 197, 199, 204
supply chain management and, 68
Competitive weapon, 178
Competitiveness, Samsung Group, 136
Competitor analysis, 197, 198, 201
Comshares Commander Executive Information System, total innovative management, 119
Concept development, Eastman Chemical, 135
Conceptual framework, macrologistics strategy, 10
Concurrent engineering, 153, 159
Nippon Steel Corporation, 76
Connectivity, supply chain management and, 61
Consensus
improvement, 140
in quality function deployment, 30
Taguchi methods and, 38
Consumer behavior, 189
Contacts, incidents and problems, 210
Continuous improvement, 22, 23, 201, 217, 233
3-M, 54
cycle of, 170
in macrologistics strategy, 13
total innovative management, 111, 112
Continuous learning, 248
Contract, 156
Control, total innovative management, 113
Control charts, 156
Control factors, Taguchi methods and, 36
Controlled integration, 166
Coordination of work, 216
Copacino, William C., 59
Core business competition, 225
Core competencies, 181, 184, 219, 248, 250, 251
in macrologistics strategy, 11
Rubbermaid, 131
total innovative management, 116
Core output, 250, 251
Core process, 220, 251
Corporate climate, total innovative management, 114
Corporate culture, 140, 141, 226

Corporate history, total innovative management, 114
Corporate measurement system, 197, 231
  balanced scorecard, 231–44
Corporate performance, total innovative management, 120
Cost, total, concept of, 243–44
Cost savings, 159, 219, 250
Council, logistics, balanced scorecard, 234–35
Coy, Peter, 116
Cram school, 140
Creative management styles, total innovative management, 119
Creative problem solving, 146
  total innovative management, 118
Creativity, 209
  circles for, total innovative management, 119
Critical success factor, 29, 193, 196, 197, 198, 200, 201, 202, 203, 207, 212, 215, 216, 219, 231
  AT&T, 200–202
  competitive marketing analysis, 197–98
  criteria for, 196
  General Electric, 208–9
  interviews, 196–97, 201
  in macrologistics strategy, 11
  measures-and-reporting sequence, 202
  Motorola, 209–10
  sources of, 198–200
  tailoring, 203–4
  total innovative management, 118
Critical thinking, 209
Cross-functional action, 180, 218, 220, 231, 236
  ARCO project, 126
  Burlington Northern, 122
  FPL Group, 44
  information exchange, 102
  streamlining, 218
  supply chain management and, 64
  synergy and, 8
  teams, 23, 144
    3-M, 51
    Rubbermaid, 131
Cross-functional quality team, in macrologistics strategy, 4
Cross-market, 193
Cross-organizational cooperation
  ARCO project, 126
  project teams, Burlington Northern, 121
Cross-training, 151
CSF. *See* Critical success factor
Culture, 140, 142, 225
Customer, data base, supplier data bases, 240
Customer-driven company, 39, 188
Customer feedback, in quality function deployment, 31
Customer focus, 145
Customer information, FPL Group, 49
Customer-interaction, 211
Customer participation, 31
Customer relations, Rubbermaid, 132
Customer relationships, 173
Customer requirements, 31, 32
  quality function deployment and, 34
Customer satisfaction
  3-M, 51
  Becton Dickinson, 74
  enterprise logistics management, 105
  FPL Group, 41, 47, 48, 50
  supplier, alliances, Becton Dickinson, 72, 74
  total, total innovative management, 111
  total innovative management, 111
Customer service, 142
Customer service automation, 207
Customer-supplier orientation, 182
Customer-supplier partnership, 160
Cyber-culture, logistics, 89
Cyberspace, 89, 94
Cycle, 149
Cycle time, 155, 159, 162, 167, 220

D

Daily control systems, FPL Group, 47
Daily management, FPL Group, 49
Dantotsu, defined, 204
Data scatter, effect of, 102–3
Davenport, Thomas, 218
Decentralization, 176
  Hitachi, 133
Decide-test-evaluate-change, 249
Defect tracking, 211
Delivery cycle, 161

Demand and supply core processes, 221
Demand process, supply process and delivery process, 220
Deming Prize, 39, 46, 49, 225
Design, robust, Taguchi methods and, 36
Differentiation, total innovative management, 118
Distribution, knowlege links, 86
Distribution resource planning
  Becton Dickinson, 72, 74
  enterprise logistics management and, 105
Downsizing, 151, 224, 229, 245
  macrologistics strategy and, 10
DRP. *See* Distribution resource planning
Drucker, Peter, 116
Dunaine, Brian, 116

E

Eastman Chemical, 134–35
EDI. *See* Electronic data interchange system
Edison Award, 39
EFT. *See* Electronic funds transfer
Electronic data interchange system, 83, 92, 154, 241
  capability of, 165
  FPL Group, 49
  information system based on
    Burlington Northern, 121
    Moorhead Malting Facility, 124
  Levi Strauss & Co., 97
Electronic funds transfer, 241
Electronic mail, 241
Electronic order taking, 151
ELM, 105
Employee participation, 150, 172, 180
Empowerment, sense of, 6, 8, 147, 153, 159, 164, 180, 217
  enterprise-wide logistical model, leadership, 239
  information technology and, 99
  Levi Strauss & Co., 97
  total innovative management, 119
Enabling activity, 248
Engineering
  adaptive, macrologistics strategy, 11
  innovative, Taguchi methods, 35
Enterprise, 183, 184
Enterprise-logistics management, 105
Enterprise-wide process, 234, 235
Environment, total innovative management, 144
Environmental factors, importance of, 199
Eureka, William, 32, 35
European Quality Award, 225
Evans, James, 28, 150
Evergreen contract, 154, 169
Execution, seamless, ARCO project, 125
Expert systems, 92–93
  total innovative management, 118
Extended enterprise, 175, 181, 194
  Burlington Northern, 121
  or Kierestu, 177
External customers, 211
  Eastman Chemical, 134
External data, 206, 207

F

Fact-based management, 227
Feedback, 149, 152, 248
Fernandez, Ricardo, 29, 155
Field benchmarking, 206
Flowchart, 206, 237
Focus groups, 132, 144, 191
Focused factory, 172
Folklore processes, 219
Followership, 170
FPL Group, 39–50, 47
  application programs, to support logistics activities, 46–47
  corporate description, 40
  Deming Prize, 44–45
  improved logistics methods, 48
  information systems, Deming Prize, 44–45
  logistics
    information systems and, 44–45
    in nuclear operations, 47–48
  policy deployment, 43–44
  profile, 40–41
  services organization, information systems and, 45–46
  training, 48–49
Freight system, total innovative management, 112

Full-factorial method, in quality function deployment, 33
Future search conference, 222
Futures technology, Eastman Chemical, 135

G

General Electric, critical success factors, 208–9
Generic cross-industry study, 205
Gleick, James, 87
Global competition, 188, 190
total innovative management, 116
Global cultures, total innovative management, 119
Global entrepreneur, 236
Global information system, 242
Global leadership, 210
Global logistics, 159
best practice, supply chain management and, 61
supply chain management and, 68
Global marketplace, 175, 236
Global network, 139, 194
Global operational strategy, 178
Global sourcing, 177, 178
Globalization, 176, 185
producer life cycle and, 88
supply chain management and, 59, 63
total innovative management, 119
Globally based suppliers, 139
Goal setting
ARCO project, 126
Burlington Northern, 123
in policy deployment, 28
total innovative management and, 127–30
Greene, Richard Tabor, 27, 31, 197
Group conferencing, 223

H

Hall, Gene, 219
Hammer, Michael, 218
Help desk, 211
Help desks, 210
Hewlett-Packard
interactive network, 190
value stream quality system, 190–91
Hiam, Alexander, 152
Higgins, James, 116, 118
High performance team, 129
Hitachi, 132–33
Horizontal integration, 176
Hoshin planning, 23, 27, 29, 37
in policy deployment, 28
House of Quality, 34
customer needs alignment, 33
Human resource, inventory, supply chain management and, 67, 68

I

I SCHMLT, 65
IBM, 142–43, 229–30
Idea-assessment system, 3-M, 52
Identity processes, 219
IKEA, 138–40
Image processing, 94–95
Imai, Masaaki, 150
Implementation, macrologistics strategies, 243–51
action plan, 251
implementation framework, 246–49
"world-class" macrologistics goals, 249–50
Improvement, continuous, in macrologistics strategy, 13
In-plant representative, 241
In-plant supplier representatives, 162
In-plant vendor representatives, 153
Incremental innovation, 137
Incremental thinkers, 138
Indicator owner/champion, 237
Industrial espionage, 205
Industrial integration, 181
Info-corporation, 94
Informal multifunctional teams, Eastman Chemical, 135
Information, quality, relationship between, 101–2
Kao Corporation
Information literacy, 182
value stream quality system, 182
Information network, innovative, Hitachi, 133
Information sharing, 167

Information systems and technology, 81–105
architecture, 98–99
bulletin boards, 90–91
collaborative networks, 91–92
data scatter, 102–3
databases, 90–91
electronic data interchange systems, 92
ELM, 105
expert systems, 92–93
image processing, 94–95
impact of technology, 82–83
information, quality, relationship between, 101–2
Internet, 91
inventory, supply chain management and, 68
Kao Corporation, 100–101
knowledge links, 85
productivity of, 85–87
Levi Strauss & Co., 97
Livonia, 103–4
local area networks, 90
logistics quality, 99–100
macrologistics, integration and, 95–96
in macrologistics strategy, 11
networks, 88–92
neural networks, 93
parallel systems, 93
product life cycles, shortening of, 87–88
satellites, and superhighways, 88–92
superhighways, 88–92
supply chain, 67
trends, emerging, 83–84
value-added networks, 92–93
virtual workspaces, 94
World-Wide Web, 91
Innovation concepts, n139, 117, 146, 192, 208, 210, 225
applied research, Eastman Chemical, 135
Eastman Chemical, 134, 135
four types of, 117–19
management quality, adaptive learning, integration of. *See* Total innovative management
Nippon Steel Corporation, 76
Rubbermaid, 131
Samsung Group, 136
Innovation research program, 146
Inplants, synergy and, 8
Instructional engineers, 209
Integrated communications, Microage Computer Centers, 78
Integrated logistics
supply cahin management, Becton Dickinson, 72
supply chain management and, 68
Integrated supply chain system, 60
Integration, 181, 185, 193, 199, 202, 217, 220, 221, 226, 232, 246. *See also* Keiretsu
macrologistics, 21–22
in macrologistics strategy, 11, 16, 20, 22, 23
supply chain management and, 65
Inter-organizational leadership team, supply chain management, 66
Inter-organizational task force, supply chain management and, 67
Interactive balanced scorecard measurement systems, total innovative management, 119
Interactive strategy, 140
Interconnectivity, 160
Interdependence driven, 182
Internal benchmarking, 205
Internal customer, 241
Eastman Chemical, 134
supplier relationships, 236
Internal data, 206, 207
Internal help desk, 211
Internal suppliers, 190
Internal transportation, 152
Internet, 91
use of, 91, 94
Intrapreneurship, total innovative management, 119
Inventory control, 146, 149, 151, 153, 157, 163, 178, 190, 211, 243
excess, ARCO project, 127
human resource, supply chain management and, 67
information system, supply chain management and, 68
just-in-time management, 171–72
impact of, 172–73

materials management, 150
technology, supply chain management and, 67, 68
total innovative management, 112, 115

J

Jamrog, Mark R., 152
Japanese Quality Movement, 155
JIT II™. *See* Just-in-Time management
Job satisfaction, 239
Johansson, Henry, 215
Joint planning, 167
Joint ventures, total innovative management, 119
Juku, 140
Just-in-Time management
Becton Dickinson, 69
Bell Labs, 169–70
Bose, 160–67
enterprise logistics management, 105
factory walk through, 152
implementation, 152–53
inventory, 171–72
impact of, 172–73
inventory control, 150
management accountant, 170–71
move to, 153–55
partners, 168
partnerships, with partners, 155–58
production, 150
rep, 162
supplier certification, 158
supplier quality management, 155, 158–59
suppliers, 162
team, 167
Toyota manufacturing system, 151
transportation process, 167–69

K

Kaizen, 150, 1159
Kamp, Ellen, 218
Kanban, 150, 151, 190
Kao Corporation, 100–101
Keen, Peter, 151, 218
Keiretsu, 21, 193
Keiretsu, in America, value stream quality system, 193–94
Kewen, Peter, 82
Kierestu, 177
King, Bob, 29
Knapp, Ellen, 82, 151
Knowledge-based system, 242, 243
total innovative management, 119
Knowledge links, productivity of, 85–87
Knowledge management, 190, 212
3-M, 52
assessment tool, 212
lateral thinking, total innovative management, 118
total innovative management, 119
Knowledge sharing, 3-M, 52

L

LAN. *See* Local area network
Lateral thinking, knowledge management, total innovative management, 118
Leadership skills, 111–12, 143, 160, 175, 223, 225, 227, 234
in macrologistics strategy, 11
Samsung Group, 136
total innovative management, 119
Leadership team, supply chain management, inter-organizational, 66
Learning organization, Taguchi methods and, 35
Level-of-availability monitoring, 220
Level-of-service, 220
Leverage, 248, 250
global capabilities, 188
Levi Strauss & Co., 97
Lindsay, William, 28
Livonia, 103–4
Local area networks, 90
Logistical concepts, supply chain management and, 63
Logistical control systems, FPL Group, 45
Logistical deployment, 191
total innovative management, 118
Logistical global strategies, total innovative management, 119
Logistical service relationships, 201

Logistics, 105, 149, 152, 165, 177, 181, 205, 215, 216, 217, 218, 220, 231, 232, 239, 240, 242, 250. *See also* Enterprise-logistics management
channel, 149
FPL Group, 39, 44
global, best practice, supply chain management and, 61
information, 241
information systems, 202
information technologies, producer life cycle and, 88
information technology, 82, 84, 95, 96
integrated, supply chain management and, 68
Livonia, 103
in macrologistics strategy, 7, 13, 15
producer life cycle, 87
supply chain management and, 59, 63
Taguchi methods and, 35, 37
total innovative management, 10–9, 110, 119
value chain quality, 78
Logistics-based critical success factors, 198
Logistics-based information, 202
Logistics-based quality control tools, 172
Logistics benchmarking, steps to, 206–7
Logistics competencies, 219
Logistics cost, 241
database, 240, 241
information, 240
Logistics council, 235, 238
balanced scorecard, 234–35
Logistics cyber-culture, 89
Logistics function, 236
Logistics improvement program, 159, 191
Logistics indicators, 236, 237
Logistics management, 151
FPL Group, 41, 46, 47
total innovative management, 109
training, 223
Logistics managers, 149
Logistics maps, Becton Dickinson, 74
Logistics measurement
system, 233, 234
team, 238
Logistics model, 203, 221
Logistics objectives, 237
Logistics planning, 191
Logistics policy, 249
Logistics priority, 238
Logistics process, 240
enhancements, 220
innovation, total innovative management, 117
total innovative management, 117
Logistics process enhancement, 220–21
Logistics quality, 99–100
Logistics strategy, 172
Logistics value chain, FPL Group, 45
Loyalty, supply chain management and, 63

M

Macro-process, Burlington Northern, 121
Macrologistics strategies
examples of, 12–13
implementation, 243–51
action plan, 251
implementation framework, 246–49
"world-class" macrologistics goals, 249–50
integration and, 95–96
overview, 5–6
quality management, integration of, 9–12
system model, 15–24
Malcolm Baldrige National Quality Award, 3, 9, 119, 220
Management by objectives, FPL Group, 43
Management by wandering around, total innovative management, 119
Management consensus, FPL Group, 44
Management control systems, 203
Management cycle, 153
total innovative management, 110, 111, 114, 115
Management design, total innovative management, 114
Management functions, total innovative management, 114
Management innovation, total innovative management, 118
Management performance, total innovative management, 115

Management planning
control systems, 203
process, comprehensive, 20
Management quality, innovation, adaptive learning, integration of. *See* Total innovative management
Management quality system, total innovative management, 119
Management resources, total innovative management, 114
Management reviews, FPL Group, 44
Management structure, decentralized, Hitachi, 133
Management style, 138
Management systems, 143
Management teams, Samsung Group, 136
Mandated processes, 219
Manufacturing resource planning, 154, 162, 171
Becton Dickinson, 72, 74
enterprise logistics management and, 105
Market development
Eastman Chemical, 135
team, Eastman Chemical, 135
Market/product innovation, 216
Market research, 205
Eastman Chemical, 135
Market share, 225
Marketing innovation, total innovative management, 117, 118
Mars, balanced scorecard, 240–42
Masterplan, 130
Material handling, 149
MaxComm, whole organization systems change approach, 221–23
MBO. *See* Management by objectives
McLaughlin, Gregory, 27
Measurement fluency, 182
value stream quality system, 183–84
Measurement monitoring system, Becton Dickinson, 72
Measurement system, 231, 235
developing, 235–39
supply chain management and, 62
total innovative management, 114
Measures, for balanced scorecard, types of, 234
Measures of innovation, Eastman Chemical, 135
Microage Computer Center, 77–78
Microage Computer Centers, logistics, 77
Miniaturation, 176
Mische, Michael, 56, 57
Mobilization, 11, 16, 20, 22, 23, 109, 112, 114, 151, 153, 159, 246, 247, 250
macrologistics, 18–21
Moorhead Malting Facility, 124
Motivational mechanism, 150
Motorola, critical success factors, 209–10
MRP. *See* Manufacturing resource planning
Multifunctional teams, Eastman Chemical, 135
Mutual improvement, 205

N

Needs identification, Eastman Chemical, 135
Network management, 211
Networked organizations, total innovative management, 119
Networking, 88–92, 169
New product
development process, Eastman Chemical, 134
teams, cross-functional, 3-M, 51
Nippon Steel Corporation, 73–77
Noise factors, Taguchi methods and, 37
Norfolk Southern, 145–47
Not-invented-here syndrome, 189, 204

O

OLIVER. *See* On-Line Interactive Visual Employee Resource
On-Line Interactive Visual Employee Resource, 97
Operational benchmarking, 201
Operational costs, 178
Operations management, 159
Order-cycle process, Microage Computer Centers, 77
Order fulfillment, 5, 9, 246, 249
goal, 10
Organizational culture, 142, 143
Samsung Group, 136, 1136

Organizational innovation, total innovative management, 120
Organizational learning, 217, 227
3-M, 52
total innovative management, 120
Organizational models, 175
Organizational processes, in macrologistics strategy, 7
Organizational reengineering, 219
Original thinking, 137
Outcome indicators, 13, 234, 236
Outsourcing, 151

P

Painter, Jim, 56
Parallel systems, 93
Parameter design, Taguchi methods and, 37
Pareto analysis, 30, 38, 227
Partnering, 60, 72, 153, 164, 183, 206, 244
Pearce, Glen Stuart, 33
Performance measurement, 196, 201, 203, 248
supply chain management and, 63, 67
synergy and, 9
Performance promise, ARCO project, 125
Pit crew, 129
Plan-do-check-act-cycle, 47, 227, 249
Platform teams, 144
Policy alignment, 29, 201
Policy deployment, 19, 23, 27, 29, 37, 38, 200
alignment and, 29
description of, 27–28
goal setting and, 28–29
Policy development, quality function deployment and, 27–54
Porter, Michael, E., 59
Possession costs, Becton Dickinson, 71
Predictive measures, 184
Priority processes, 219
Problem solving, 122, 209, 249
creative, total innovative management, 118
Process, defined, 215–16
Process blindness, 216
Process capability, 156
Process control charts, total innovative management, 113
Process enhancement, Burlington Northern, 121
Process facilitation, Eastman Chemical, 135
Process improvement, 239
Process indicators, 236
Process Information Technology, 181
Process investment, 218
Process management, 215–30
IBM, 229–30
MaxComm, whole organizational systems approach, 221–23
Xerox, 225–26, 227
Process mapping, 77, 158, 182, 206, 208, 221
Process movement, 217
Process orientation, 182
value stream quality system, 182–83
Process paradox, 219
Process redesign, 207, 215, 248
teams, 3-M, 51
total innovative management, 119
Process reengineering, 215–30, 217–18
Procurement cycle, 183
Product innovation, 117, 118, 134, 225, 229
Product life cycle, 190
shortening of, 87–88
Production scheduling, 163
Production system, 151
Productivity improvements, 179
Profit-center approach, 162
Project completion cycles, 135
Prosumers, knowlege links, 86
PS process, 75
Pull system, 172
Purchasing/supplier, quality management, 38
Push process, 151

Q

QFD. *See* Quality Function Deployment; Quality function deployment
QIDW. *See* Quality in Daily Work
Quality control, 37, 42, 43, 44, 49, 113, 136, 143, 145, 146, 149, 152, 156, 172, 176, 178, 179, 182, 184, 188, 189, 212, 215, 218, 219, 227, 233, 234, 239, 249
circles, 42
control, 152
management by facts, 204

teams, 211
total innovative management, 112
Quality council, 180, 185
Quality function deployment, 18, 23, 27, 29, 30, 31, 32, 37, 38, 39, 40, 44, 49, 201
overview, 29–33
policy development and, 27–54
Taguchi methods, 33–37
Quality in daily work, 45, 46
Quality in manufacturing, 177
Quality information systems, 103
Quality inspections, 181
Quality learning organization, 227
Quality loss function, Taguchi methods and, 36
Quality management, macrologistics strategies, integration of, 9–12
Quality principles, 175
Quality product, 158
Quality rating system, 164
Quality tools, 183, 191
Quality vendor, 158

R

Radical innovation, 137
Radical thinkers, 138
Ramanujam, Vasudevan, 118
Razor-sharp execution, 129
Razor-sharp goal, 126, 129, 130
Reengineering, 119, 151, 211, 212, 217, 219, 220, 221, 223, 239, 249
whole systems change and, 221
Regional stocking location/distribution center network, 221
Relationship
with customer, 139
with suppliers, 139
Reliability studies, 156
Replenishment, 221
Reporting structure, ARCO project, 125
Resource planning
distribution, enterprise logistics management and, 105
manufacturing, enterprise logistics management, 105
Resource sharing, 221
Responsiveness, 209, 229, 236
Restructuring, total innovative management, 119
Return on quality, 191
Reverse logistics, in macrologistics strategy, 12
Reward, recognition systems, Becton Dickinson, 72
Risk management
Becton Dickinson, 71
Microage Computer Centers, 77
Risktaking, 136
Root cause analysis, 185
Rosenthal, Jim, 219
Ross, Joel, 151
RSV. *See* Regional stocking location
Rubbermaid, 130–32
Ryan, Nancy, 32, 35

S

Safety management, 146, 236, 242
Sales force, 144, 190, 201
training, 190
Samsung Group, 136–37
Satisfaction, of customer. *See* Customer satisfaction
Scenario forecasting, total innovative management, 119
Schonberger, Richard, 150
Scientific method, 227
SCM. *See* Supply chain management
Scorecards, balanced, 22
Self-Directed Work Teams, 180
Self-management, 119, 169
Serafin, Raymond, 116
Seven S's, 225
Shared information systems, Becton Dickinson, 73
Shared values, Becton Dickinson, 74
Shareholder value, 142
Sheridan, Bruce, 29
Shiseido, 140–41
Signal-to-noise ratio, Taguchi methods and, 37
Silo mentality, supply chain management and, 64
Six-sigma defect reduction, 248
Small-scale teams, Burlington Northern, 121

Speed strategies, 186, 218, 245, 250, 251
in macrologistics strategy, 6, 22
total innovative management, 119
Sprow, Eugene, 33
Stakeholders, ARCO project, 125
Stata, Ray, 113, 117
Statistical process control, 29, 47, 49, 156, 164, 172, 173
Statistical tools, 227
Stranco, 137–38
Strassmann, Paul, 217
Strategic alliance, 114, 119, 183, 225, 241
Strategic benchmarking, 201, 205
Strategic partnership, 247
Strategic planning, 191, 199, 207, 212, 216, 218, 225, 229, 233, 245, 246, 248, 249, 250
Stretch, concept of, 6, 7, 28, 245, 250, 251
Stretch goals, added value and, 6–8
Structured executive planning, 223
Structured project management system, Eastman Chemical, 135
Success factors, 202
critical, in macrologistics strategy, 11
Supplier, 8, 58, 72, 97, 139, 149, 153, 154, 155, 156, 157, 162, 171, 172, 180, 181, 183, 221, 240, 244, 247, 250
customer, alliances, 72, 74
data bases, 240
in macrologistics strategy, 12
quality function deployment and, 33
quality management, 38
synergy and, 8
Supplier base, 243
Supplier certification, 155
just-in-time management, 158
Supplier-customer relationship, 184
Supplier information, FPL Group, 49
Supplier partnering, 23
Supplier policy deployment, 158
Supplier quality, 155, 159
Supplier quality management, 158–59, 159, 160
just-in-time management, 155, 158–59
Supplier requirements, ARCO project, 125
Supplier selection, 155
Supply chain management
connectivity, 61–62, 66
defined, 55–56
facilitation, 65–68
implementation of, 60–64
information system, 67
initiation, e and to facilitate, 65–68
integration, 240
inter-organizational support systems, 62–63, 67
leadership team, inter-organizational, 66
in macrologistics strategy, 19
need for, 56–60
resources, sharing of, 63–64, 67–68
total innovative management, 117
Supply chain operations, 237
Supply chain process mapping, 65
Supply core process, 220
Support, strategic role of, 210–11
Synchronizing values, 246
Synergy, new sources of, 8–9
System integrators, 139
System model, macrologistics, 15–24
Systemic creative thinking, 137
Systems perspective, overview, 183

T

Taguchi methods, 33, 35, 36
case studies, 35
design of experiments, 49
quality function deployment and, 33–37
Target stocking levels, 220
Task force, inter-organizational, supply chain management and, 67
Team charter, ARCO project, 125
Team member responsibilities, ARCO project, 125
Team organization, ARCO project, 125
Teamwork, 170, 223, 244, 250
Technology, 81–105
architecture, 98–99
bulletin boards, 90–91
collaborative networks, 91–92
data scatter, 102–3
electronic data interchange systems, 92
ELM, 105
expert systems, 92–93

image processing, 94–95
impact of technology, 82–83
information, quality, relationship between, 101–2
Internet, 91
Kao Corporation, 100–101
knowledge links, productivity of, 85–87
Levi Strauss & Co., 97
Livonia, 103–4
local area networks, 90
logistics quality, 99–100
macrologistics, integration and, 95–96
networks, 88–92
parallel systems, 93
product life cycles, shortening of, 87–88
trends, emerging, 83–84
virtual workspaces, 94
Technology center, 144
Technology fusion, Hitachi, 133
Technology inventory, supply chain management and, 67, 68
Technology transfer, Eastman Chemical, 134
Temporal factors, 199
Texaco, value stream quality system, 192–93
Third-party logistics support, 247
Thor, Carl, 204
3M, 50–54
Time boundedness, 182
Toffler, Alvin, 86, 87
Total cost, 243
Becton Dickinson, 71
breakthrough, 244
concept of, 243–44
Total customer satisfaction, 111
Total innovative management, 109–20, 112–13, 119
action verbs, 127–30
Burlington Northern, 121–30
ARCO project, 125–27
background information, 121–22
Moorhead malting facility, 124–25
small-scale teams, 122–23
urgent measurable goals, 123
Chrysler Corporation, 143.145
chimneys, 143–44
culture, changes in, 144–45
leadership style, changes in, 144–45
macro structure, changes in, 144–45
management systems, changes in, 144–45
platform teams, 143–44
concept of, 109, 110, 111, 112, 113, 114, 119
control, 113
Eastman Chemical, 134–35
goal setting, 127–30
Hitachi, 132–33
IBM, 142–43
IKEA, 138–40
innovation, 117
four types of, 117–19
leadership and, 111–12
logistics
productivity and, 116
systems, application to, 115
Norfolk Southern, 145–47
Rubbermaid, 130–32
Samsung Group, 136–37
Shiseido, 140–41
Stranco, 137–38
as system, 114–19
Total quality management, 3, 4, 5, 9, 10, 13, 19, 42, 103, 104, 110, 111, 160, 212, 249
3-M, 51
failures, 217, 219
Total supply chain, supply chain management and, 59, 63
Traffic management system, 240
Training, 172, 182, 185, 217
Samsung Group, 137
Transformation, 221, 245
Samsung Group, 136
Transformational leadership
3-M, 53
total innovative management, 119
Transportation process, just-in-time management, 167–69
Tribus, Myron, 110
Trouble call management system, FPL Group, 49
Two-hats principle, 156

U

Up-stream control measures, predicition, 234
Urgent measurable goals, 123
Usage costs, Becton Dickinson, 71
User's satisfaction, total innovative management, 111

V

Value-added networks, 177, 184, 243, 244, 248
  technology and, 93
Value-added producer life cycle, 87
Value-added quality, in macrologistics strategy, 6
Value chain, 27, 29, 121, 127, 140, 149, 153, 158, 194, 198, 202, 215, 220, 236, 239
  quality, ten-point model, 78–79
Value constellation, 140
ValueStream Quality System, 175–94
  adding value, manufacturing, 177–78
  competitive advantage, loss of, 178–80
  customer-supplier/stakeholder orientation, 184–87
  evolution, of value system, 176–77
  Hewlett-Packard, 190–91
  information literacy, 182
  interdependence driven, 183
  Keiretsu, in America, 193–94
  measurement fluency, 183–84
  process orientation, 182–83
  stakeholder value, 176
  Texaco, 192–93
  value proposition, 180–82
  Whirlpool, 188–90
Value synchronization, 246
Varadarajan, Rajan, 118
Vendor, 229, 240, 241
  excellent, 158
  management of, 220
  selection process, 183
  systems, 185
Vengrow, Jeffrey, 175
Vertical integration, 176, 205
Virtual factory, 183
Virtual workspaces, 94
Voehl, Frank, 55, 177, 221
Voice of customer, 201, 250
Vonchek, Arthur, 60

W

Wade, Judy, 219
Wage and price controls, 212
Waller, Alan, 56
Warehouse system, total innovative management, 112
Waste, 150, 151, 180, 181
Whetham, C. D., 113
Whirlpool, ValueStream quality system, 188–90
WIP. *See* Work in process
Woodruff, David, 116
Work in process, 172
Workflow, 215
Workout, 208
World-Class macrologistics goals, 249–50
World -Class quality, 5, 22, 23, 69, 105, 155, 189, 248, 249, 250, 251
  total innovative management, 118
World Class supply chain management, 73
Worth, concept of, 218

X

Xerox, 225–26, 227

Y

Yahagi, Seiichiro, 115

For Product Safety Concerns and Information please contact our EU
[illegible]
[illegible]

Printed and bound by CPI Group (UK) Ltd, Croydon, CR0 4YY

For Product Safety Concerns and Information please contact our EU representative GPSR@taylorandfrancis.com Taylor & Francis Verlag GmbH, Kaufingerstraße 24, 80331 München, Germany

Printed and bound by CPI Group (UK) Ltd, Croydon, CR0 4YY
08/05/2025
01864362-0009